P9-BZJ-685

Faust in Copenhagen

ALSO BY GINO SEGRÈ

A Matter of Degrees

GINO SEGRÈ

Faust in Copenhagen

A Struggle for the Soul of Physics

VIKING

VIKING
Published by the Penguin Group
Penguin Group (USA) Inc., 375 Hudson Street
New York, New York 10014, U.S.A.
Penguin Group (Canada), 90 Eglinton Avenue East, Suite 700, Toronto,
Ontario, Canada M4P 2Y3 (a division of Pearson Penguin Canada Inc.)
Penguin Books Ltd, 80 Strand, London WC2R 0RL, England
Penguin Ireland, 25 St. Stephen's Green, Dublin 2, Ireland (a division of Penguin Books Ltd)
Penguin Books Australia Ltd, 250 Camberwell Road, Camberwell, Victoria 3124, Australia
(a division of Pearson Australia Group Pty Ltd)
Penguin Books India Pvt Ltd, 11 Community Centre, Panchsheel Park, New Delhi–10 017, India
Penguin Group (NZ), 67 Apollo Drive, Rosedale, North Shore 0745, Auckland,
New Zealand (a division of Pearson New Zealand Ltd.)
Penguin Books (South Africa) (Pty) Ltd, 24 Sturdee Avenue, Rosebank,
Johannesburg 2196, South Africa

Penguin Books Ltd, Registered Offices: 80 Strand, London WC2R 0RL, England

First published in 2007 by Viking Penguin, a member of Penguin Group (USA) Inc.

1 3 5 7 9 10 8 6 4 2

Excerpts from *Collected Works of Niels Bohr,* series edited by E. Rudinger. By permission of Niels Bohr Archive, Copenhagen and Elsevier.

Excerpts from *My World Line* by George Gamow (Viking, 1970). Excerpts and drawings from *Thirty Years That Shook Physics* by George Gamow, illustrated by the author (Doubleday, 1966). By permission of Igor Garnow.

Excerpts from *Scientific Correspondence with Bohr, Einstein, Heisenberg and Others* by Wolfgang Pauli. By permission of Springer Verlag, Heidelberg.

Photograph credits: Number 1: Photo: Kavaler / Art Resource, New York; 2, 5, 6, 11, 12, 13, 14, 18: Niels Bohr Archive: 3, 4, 7, 9, 15: AIP Emilio Segrè Visual Archives; 8: Photograph by Benjamin Couprie, Institut International de Physique Solvay, courtesy AIP Emilio Segrè Visual Archives; 10: AIP Emilio Segrè Visual Archives, Goudsmit Collection; 16: Photograph by P. Ehrenfest, Jr., courtesy AIP Emilio Segrè Visual Archives, Weisskopf Collection; 17: Photograph by Erik Gustafson, courtesy AIP Emilio Segrè Visual Archives, Margrethe Bohr Collection; 19: Photograph by A. John Coleman, courtesy AIP Emilio Segrè Visual Archives, Physics Today Collection.

ISBN 978-0-670-03858-9

Printed in the United States of America
Set in Electra and Cloister Designed by Francesca Belanger

In memory of George Gamow

Contents

Introduction

THIS IS A BOOK about seven physicists, six men and one woman, who attended a small annual gathering in Copenhagen in April 1932. To be honest, only six of them were actually there. The seventh, Wolfgang Pauli, had originally intended to go, as he had in earlier years and would do so again, but he decided that spring instead to take a vacation. He was there in spirit, as you will see.

Four of the seven—Niels Bohr, Paul Dirac, Werner Heisenberg, and Wolfgang Pauli—would be placed in most physicists' selection of the century's top ten physicists. Lise Meitner, the only woman in the group, ranks high on anyone's list of the century's most important experimentalists. Another of the seven, Max Delbrück, changed fields soon after the meeting, though he never stopped defining himself as a physicist. He went on to become one of the founding fathers of modern molecular biology and ranks as one of that discipline's top ten. All of them taught and mentored a generation of future scientists. The last of the seven, Paul Ehrenfest, was perhaps the greatest teacher of them all.

Physics was fortunate to have at one moment a remarkable number of individuals to help create and shape the great revolution in science called quantum mechanics. Indeed, one could say that the revolution occurred because of them. It developed very differently from relativity, the twentieth century's other major departure from physics' past. Relativity, in the special theory of 1905 and the general theory of 1916, was the work of a single individual, Albert Einstein. Both the special theory and the general theory were essentially complete in Einstein's initial formulations, requiring no revisions or subsequent interpretations. Quantum mechanics, on the other hand, emerged in 1925–26 only after a long buildup. Its details evolved over time, and its meaning continued to be debated for years. Unlike relativity, it was the work of many who struggled together, often arguing with one another as they hammered out the theory's conclusions. Its final version, the so-called Copenhagen interpretation, was contested even by some of the creators of the revolution. The questioning has not ceased.

Together Pauli, Heisenberg, Dirac, and the others created something remarkable, something that has changed all our lives in a practical sense more than any other twentieth-century scientific upheaval has. The inventions it led to, such as the transistor and the laser, are both implements that affect our daily activities and tools for future research.

This is, however, a book about the human side of science, describing not only what these physicists did but how they did it and what they were like. It focuses on these seven, though others will play a role too. There's no single answer to the question of what they were like, since physicists are as different from one another as any other group. Among them we find the gregarious and the withdrawn, the philanderers and the faithful, the rooted and the wanderers. Some were abstemious and others drank too much. There were perhaps a disproportionate number of music lovers and mountain climbers among them, but that may be because they had been told these are things physicists do. Their working habits differed: some preferred the early morning and others the late night. Some always worked alone and others required discussions with their peers. But the founders of quantum theory had one thing in common: they were geniuses at the particular calling known as theoretical physics.

They had a second common trait, perhaps not independent of theoretical-physics genius. Three of the scientists, all born between 1900 and 1902, stand out for their precocity: by their midtwenties Pauli, Heisenberg, and Dirac were leaders in the founding of the field. Several older theoretical physicists, notably Bohr, Einstein, Max Born, and Erwin Schrödinger, also played extremely important roles in the revolution, but the youthfulness of its major participants remains a striking feature. All of them had revealed their powers and were famous in the field by the time they were thirty. More accurately, all but one had achieved great prominence by then, and that one, Schrödinger, may have simply been delayed in his intellectual flowering because his years between ages twenty-five and thirty were taken up by military service during World War I.

Among all these physicists, one stands out for his personal impact on the field and on the others, not simply for his thought or achievements. In his obituary for Bohr, Heisenberg wrote, "Bohr's influence on the physics and the physicists of our century was stronger than that of anyone else, even than that of Albert Einstein." This seems, at first sight, to be the sort of generous testimonial one makes in obituaries, but Heisenberg was not given to exaggeration. I was surprised to read it, for my generation does not think of Bohr this way (I confess to being a theoretical physicist myself, though hardly in the range of geniuses). But the more I delved into the matter, the more I came to understand its truth. It is not based on relative intellectual contributions, for Einstein's were certainly far greater than Bohr's. It revolves around how Bohr's style affected the way physicists think and work, how they individually and collectively strive for answers, how they relate to their mentors, their peers, and their students. In the process of wielding this power, Bohr also became the most loved theoretical physicist of the twentieth century.

Yes, loved. Respect and admiration were feelings young physicists had for all of these greats, but love is something different. Yet it is a term that appears again and again in memoirs when physicists speak of Bohr. Part of this book's scope is to explain why this is so, what it was about Bohr's persona, his behavior, his way of thinking and working, that led others to regard him with such warmth.

One of the many factors that contributed to this affection was Bohr's lack of pretension or pomp. There wasn't a trace of personal ambition or aggrandizement in him, though few of these geniuses seem to have had much need of this. They were all secure in the knowledge of their own stature, but Bohr had an almost childlike innocence about such matters. He also worked tirelessly to improve others' work and lives. There are countless examples of people owing him their positions, their careers, and sometimes their very survival. He had a sense of who was in need, when and how to intervene, and how to make a difference.

But combined with all this was something else that seems to have played a role. Being connected to others, or as Bohr's biographer Abraham Pais calls it, being *conjoined,* was a need for Bohr, almost a necessity. His discussions were carried out as a kind of Socratic dialogue in which he slowly shaped and molded his thoughts, so much so that some said he was a philosopher, not a physicist. Bohr also loved paradoxes, believing that seeing the many sides of a problem was the way to reach resolution and clarity. His close friend Einstein described him as uttering "his opinions like one who perpetually casts about, and never like one who believes he holds the whole defining truth." This astute remark captures much of Bohr's essence—he constantly strove for that defining truth.

Bohr's need for others, for conjointness, was displayed in his relaxation as well, be that skiing, sailing, simply walking, playing a game, or going to the movies. He made others feel needed because he did need them. Bohr was certainly a great man and without guile, but his constant engaging of those he was involved with was a major factor in creating the love they felt for him.

Another of these theoretical physicists was also much loved, though his peers would use that word guardedly. Later generations find the love for him to be even more puzzling than that for Bohr. He forms an interesting contrast to Bohr because while Bohr was invariably polite, Wolfgang Pauli was invariably rude. His insults and his aphorisms became legend, but part of the legend lies in the realization that these insults were directed without regard for rank or age. Einstein, Bohr, or Heisenberg might just as easily be affronted as a student. In doing so, he never meant to hurt others. As the well-known physicist Victor Weisskopf said,

"Pauli possessed an almost childlike honesty, and always expressed his true thought directly," adding that once you became used to his style, it was easy to live with him.

Pauli might insult you, but he never ignored you, and the biting remarks directed at you became a kind of badge of honor, remembered and told to friends. With a gift for the bon mot, Pauli was often very funny in his insults. Only he would describe someone as "so young and already so unknown." But his expressive language communicated true feeling and commitment. More than thirty years after a meeting they had at the height of the quantum mechanics revolution, Bohr recalled a characteristic exchange for a historian: "I met Pauli who expressed the strongest dissatisfaction with my *treason*, and in his emotional way, which we all treasured so highly, deplored that a new *heresy* should be introduced into atomic physics."

The athletic Bohr loved to go for long hikes, ski, and chop wood, while portly Pauli preferred cabarets, nightlife, and good wine. Bohr had six sons to whom he was devoted, while Pauli had no children. Bohr, while still being deeply attached to his native Denmark, worked tirelessly after World War II for world peace and disarmament. Pauli had no interest in world affairs. Living in quintessentially neutral Switzerland, he became a symbol of physics research unsullied by worldly concerns. Pauli certainly had more warts and blemishes than Bohr, but as we know, love is not only directed toward the pure. Weisskopf, who knew all of the physics geniuses of that era well, kept a photo of Pauli on his desk, while acknowledging that Bohr was his "intellectual father."

The contrast between the two, the affection felt for both of them, and the affection they felt for each other, is manifest in a skit put on by the young physicists at the April 1932 Copenhagen meeting. That year was the hundredth anniversary of the death of Johann Wolfgang von Goethe, the passing of the man, both humanist and scientist, widely regarded as the last true universal genius. As commemorations marking the occasion took place all over Europe, this small band of physicists at the annual informal gathering decided to have a celebration of their own. It took the form of a sketch, a tongue-in-cheek adaptation to the world of physics of *Faust,* Goethe's great drama. In the script, written primarily by Delbrück, noble Bohr was identified as the Lord, sardonic Pauli as Mephistopheles, and troubled Ehrenfest as Faust. As in Goethe's version Mephistopheles has the wittiest lines, but that was of course true of Pauli's real-life speech as well.

The skit, meant as comic relief from the intensity of the week's discussions, remains a fascinating portrayal of the world of physics seen through the eyes of its very young practitioners. They were the writers and producers of the parody as well as its actors. Though affectionately mocking their distinguished elders, many only a few years older than they were, these young physicists knew all too well that Bohr, Dirac, Heisenberg, and Pauli had made lasting contributions to their field by the time they were little older than twenty-five. They also remembered the warning uttered by the student Baccalaureus in Goethe's *Faust*

> **When more than thirty years are told,**
> **As good as dead one is indeed:**
>
> (*Faust, Part II,* act 2, 222–23)

and worried about their own immediate future.

The year of the meeting was a pivotal one for them. The midsummer detection of the positron, the electron's antimatter companion, marked, as we shall see, the joining of special relativity to quantum mechanics. This completed for the community, with a few remarkable exceptions, the experimental confirmation of what is still the century's most profound and far-reaching physics revolution.

On the other hand, the discovery just before the meeting of the neutron and, a few months later, the first experimentally induced nuclear disintegration ushered in another revolution in physics, introducing us to the era of nuclear physics. Its effects on our worldview and on mankind's potential for destruction still hover over us.

The year also saw the beginning of research with the cyclotron, signaling the transition in physics research from small science to big science. Whereas a single individual, James Chadwick, had discovered the neutron, efforts at the cyclotron required a team of experts and considerable financial resources. Large-scale experiments now became common. Only seven years after the meeting, a skeptical Bohr would comment that the fissionable material for a nuclear weapon could only be obtained by, metaphorically speaking, turning "the United States into one huge factory." And of course that happened.

Those discoveries of 1932, sometimes called the Miracle Year of experimental physics, also shifted the emphasis in physics from theory to experiment, from research done with pencil and paper to research done with sophisticated tools in a laboratory. The two modes of working inevitably go hand in hand, but there are times when one takes center stage and times when the other does. While theory's advances in understanding quantum physics had dominated the decade before the meeting, advances achieved in the laboratory marked the immediate period after it.

By concluding with the neutron's discovery, the Copenhagen skit points to this shift. It also eerily prefigures many of the personal problems the physicists, young and old, would encounter in the years to come. With hindsight, we see what a watershed 1932 was for them. Prior to it, they were a small community, the only tension among them induced by who would be the first to reach commonly pursued goals. They worked, ate, and traveled together, swam, played music, climbed mountains. Above all the physicists talked endlessly to one another, occasionally as rivals, but only in an intellectual sense because, in the end, they were friends and comrades. That congeniality was shattered by the ascent to power in Germany of Adolf Hitler in January 1933.

Though none of the seven physicists who are this story's focus was religiously observant, four of them, including Bohr, were at least part Jewish. In 1933 they had to begin worrying about personal safety and emigration. By little more than a decade later many of that small physics community found themselves pitted against one another in a deadly battle, thrust into the making of Faustian bargains they could not have contemplated a few years earlier.

There is a good deal of science in the book, as much as can be comfortably accommodated without the machinery of equations, the accepted language of theoretical physics. I have tried to describe the concepts accurately within these limits, hoping not to overly burden the curious reader while still challenging him or her. The book's ultimate emphasis is on the way this group of seven and their immediate community dealt with one another and with their own particular demons. To achieve this, I have looked back in time to the dozen years that preceded the meeting to see how the individuals shaped their beliefs and personalities and forward to trace the immediate consequences of their actions.

I won't deny that I am prejudiced by a lifetime in the physics profession, but these individuals were true titans. We often do not know what of recent science will stand the test of time, but the contributions of these people will last forever. Part of their success was due to their being young and enthusiastic at a crucial moment, the dawn of quantum mechanics and of nuclear physics, but they seized that moment and shaped the field. Their work was the product of an ensemble: one of them was more original, another more critical, and yet another more daring. Together they created a magical instant in history. Hundreds of years from now, their names may only be footnotes in science textbooks, but their work will continue to shape the way our descendants think.

This singular time, epitomized by the few individuals gathering for a weeklong meeting in a Copenhagen room in April 1932, is the story's hinge. I would like to introduce it by telling you about how and why I started writing it.

Chapter 1

Munich Then and Now

Thus let creation's ample sphere
Forthwith in this our narrow booth appear,
And with considerate speed, through fancy's spell,
Journey from heaven, thence through the world, to hell!

Goethe, *Faust,* Prologue, 240–43

A FEW YEARS AGO I flew to Munich, Germany, to attend an international conference on neutrino physics, my research specialty. Since I hadn't been able to leave until the night before the conference started, my flight to Europe landed just as the conferees were heading for the meeting. A train from the airport to downtown, a quick shower, shave, and breakfast in my hotel, and I was off to join them, but I felt some fresh air would do me good before I faced four or five hundred neutrino cognoscenti. The conference was being held at the Technical University, and the forty-minute hike there through pleasant downtown Munich would be just what I needed. It was a beautiful mid May morning, and the brisk walk would help clear my thoughts, preparing me for the complex discussions ahead.

Uppermost in my mind that morning was Wolfgang Pauli's suggestion, made more than seventy years earlier, that this unlikely particle, the neutrino, exists. In the 1920s matter was assumed to consist entirely of electrons and protons. It seemed like science fiction to postulate the existence of an additional entity, one with no electric charge, almost no mass, and capable of traversing unscathed almost any measuring apparatus. By now physicists have identified many types of elementary particles and given them esoteric names, such as quark, lepton, and gluon. There are so many that the introduction of a new one is commonplace, but in 1930 it was a major step, very un-Pauli-like. He was conservative by nature in his approach to physics, not given to conjectures that lacked solid experimental support.

In some ways Pauli's neutrino was more the kind of proposal one expected from his university friend Werner Heisenberg. Though many thought Pauli's intellectual gifts were greater than Heisenberg's, he was not as daring and not as much of a bold innovator. A decade after Pauli's death, Heisenberg made a comparison between the two of them:

> Pauli's character was different from mine. He was much more critical and tried to do two things at once. I, on the other hand, thought this is really too difficult, even for the best physicist. He tried, first of all, to find inspiration in the experiments and to see, in a kind of intuitive way, how things are connected. At the same time, he tried to rationalize his intuitions and to find a rigorous mathematical scheme, so that he could prove everything he asserted. Now that is, I think, just too much.

Heisenberg was right in his assessment. Pauli's approach to physics was more methodical. But he did abandon his normal caution this once, when he proposed the neutrino's existence. There was no "rigorous mathematical scheme" into which he could fit the mysterious particle, and there wouldn't be one for several years. Admittedly, by discussing his idea informally and never publishing it, Pauli stayed true to his high scientific standards. However, he knew that his reputation was such that any suggestion of his would be taken seriously, published or not.

Walking the Munich streets on that first day of my meeting, I wondered how Pauli would have viewed the thousands of us in the scientific community now working on neutrino physics. Some of us, myself included, are theorists. We go into our offices in the morning, study experimental data, examine our theories' equations, and talk to our students and collaborators, all the while trying to explain the curious behavior of these elusive particles. Our experimental counterparts are spread across the world, sometimes at giant accelerators but often in deep underground cavities within mines or under mountains. They are there because the best way to observe neutrinos is to shield the measuring apparatus from stray unwanted signals, and nothing works as well as a mile of rock. Every year or two the tribe of neutrino physicists comes together to compare notes and discuss its future plans. That is what brought us all to Munich.

Jet-lagged and headachy, my thoughts drifted to how Munich must have appeared to Pauli when he came there from his native Vienna in the fall of 1918. He was eighteen, and the pursuit of science led him to Bavaria's capital. Vienna was on the decline so that, after giving the matter some thought, Pauli decided the University of Munich would be the best place at which to study, particularly if he did so under the tutelage of Arnold Sommerfeld, its senior professor of theoretical physics and an acknowledged leader in the growing field of quantum physics.

It was a far simpler time for experimental physics. There were no huge laboratories, no need for electronics or computing specialists, no multimillion-dollar pieces of apparatus or struggles to find funding. Theoretical physics was still very much the way it is now, a solitary pursuit for most while for some a collaborative process of two or three individuals working together. For both theorists and experimentalists, the main training began, as it does today, with an apprenticeship to an older master who assigns problems to work on, sometimes still unsolved ones, and teaches the fine points of how to approach them.

Pauli's master, Arnold Sommerfeld, was the one person in the world toward whom Pauli always behaved respectfully. In later years colleagues would smile as they saw the often sarcastic Pauli deferentially saying, as he and Sommerfeld discussed an issue, "Yes, *Herr Geheimrat*, yes that is most interesting, although perhaps I would prefer . . ." This great teacher had very quickly recognized his young prodigy's talents, encouraged him, and helped launch his career. Sommerfeld steered young Pauli toward quantum theory, his own field of research at the time, but he also asked Pauli to prepare for publication an encyclopedic review article on relativity theory. It turned out to be very influential and much admired. As Einstein said of Pauli's review,

> No one studying this mature, grandly conceived work would believe that the author is a man of twenty-one. One wonders what to admire most, the psychological understanding for the development of ideas, the profound physical insight, the capacity for lucid, systematic presentation, the knowledge of the literature, the complete treatment of the subject matter, or the sureness of critical appraisal.

Three years later, by then focusing his full attention on quantum theory, Pauli discovered the *exclusion principle*, a key twentieth-century scientific idea, one whose implications affect everything from the smallest to the largest formations in our world. Why do atoms have the beautiful structure we call the periodic table of elements? Why don't all stars collapse into black holes once they have exhausted their nuclear fuel? The answer to both questions lies in the rules imposed on matter by Pauli's exclusion principle, the organizing tenet of subatomic constituents.

Pauli was truly a phenomenon.

The Munich he first saw immediately after World War I must have been much like today's handsome city. It was beautiful then and still is. Having escaped the kind of carpet bombing that destroyed much of Dresden, Hamburg, Cologne, and other major German cities during World War II, present-day Munich has retained its elegant streets lined with graceful old Baroque buildings, fin de siècle apartments, and elegant *Café-Konditoreis*, together with its beautiful park laid out in the early nineteenth century, the English Gardens.

These musings led me to more personal perspectives. In the ongoing struggle to make sense of our lives, we sometimes have moments when pieces from a distant past realign themselves and a previously unnoticed pattern emerges. Nothing has changed except perhaps the angle from which we look at those events, but the new vista suggests another meaning or even a connection of which we were previously unaware. I don't understand how it all works, but I had one of those moments on that May morning. A set of memories came rushing in as I realized that, like Pauli, my mother, Katia, also arrived in Munich in the fall of 1918 for the first time. She too was eighteen, only a few months older than Pauli, barely old enough to have been born in the nineteenth century rather than the twentieth.

Katia grew up in Cologne, the Rhineland's largest city. She was the eldest of six children in a Catholic German working-class family. Living in a slum, they faced great hardships, made even worse for her when, at age nine, she had to deal with her own mother's death from tuberculosis. At fourteen, as World War I was starting, she went to work in a factory, sustaining herself with the dream that one day she would be an artist. As soon as the war was over—November 1918—and having read in a book that artists lived in Munich, she got on a train and went there, hoping to realize her own dream of becoming one.

Pauli came to Munich to study science, Katia to study art. A few minutes earlier I had been thinking of an eighteen-year-old prodigy's entry into the world of scientific research, and now I was trying to imagine what eighteen-year-old Katia thought of her newfound freedom. How did those beautiful wide Munich streets appear to her? What did she think of the two great art museums, the Alte Pinakothek and the Neue Pinakothek, both only a few blocks away from the Technical University?

My wife's mother, Kate, also came to Munich in the fall of 1918, but neither for science nor art. Eighteen, just like Pauli and Katia, she simply came to see a friend. That friend took her along on a hospital visit, hoping her presence would cheer up an acquaintance recovering from war wounds. Kate, a northern German Jew, fell in love and eventually married the man she met that day in the hospital, lived happily with him in Munich, and had three children. Kate's husband was Bavarian, a devout Catholic, a musician, and a musicologist, but none of that mattered on June 30, 1934. The newly empowered Nazi regime seized and murdered him in the Munich suburb of Dachau. More than seventy others suffered a similar fate on that date and the next in what has since been called the Night of the Long Knives, an ignominious event in German history, the end of the protection by law of ordinary German citizens. Many of those killed were Nazi Party members perceived as potential rivals of Hitler's, but others, such as Kate's husband, were simply victims of indiscriminate Nazi assassinations. Kate eventually remarried and had another child who became my wife, Bettina, but Kate's life remained marked by the tragedy of that night.

The person responsible for the murder of Kate's husband also arrived in Munich in the fall of 1918. He was the embittered World War I veteran we know as Adolf Hitler. Coming not for science, art, or friendship, he found in the city a soil into which he could plant roots of hatred and see them grow.

In that fateful November of 1918, as Pauli, Hitler, Katia, and Kate were converging on Munich, Germany was in chaos. On the second of November, with the armed forces in retreat, a group of sailors revolted after being ordered into a hopeless battle against the Allied fleet. Their mutiny led to uprisings throughout Germany, triggering fears among many of a leftist revolution, the rising of the masses envisioned by Karl Marx.

With Germany in disarray, Munich, the capital of the state of Bavaria, experienced its own particular form of confusion. During an early November peace demonstration, the Prussian Jewish intellectual Kurt Eisner, founder of the Bavarian Independent Social Democratic Party, called for a socialist republic. Miraculously, or so it seemed to Munich's citizenry, Bavaria's king abdicated the next day, suddenly thrusting Eisner, now the unlikely leader of conservative Catholic Bavaria, into becoming prime minister of the new Bavarian Republic. It didn't last: a right-wing anti-Semite shot and killed him a few months later. Another socialist government followed briefly and then, with the local economy collapsing, a short-lived Communist one.

Deciding it was time to stop what it viewed as anarchy in the provinces, the central government in Berlin ordered troops to Bavaria. With the army in disarray after World War I, many of the soldiers now belonged to a privately supported paramilitary organization. In an orgy of killing and violence, the troops left more than a thousand dead on the streets as they entered Munich. A moderate socialist government took over in Bavaria, but that was overthrown in March 1920 and replaced by a right-wing one. Munich was now firmly established as a breeding ground for future fascist leaders, several of them spawned by those very same paramilitary organizations that had invaded the city.

During the chaos of the early 1920s, Hitler took control of the tiny German Workers' Party and began to organize it in his own image: ruthless, fanatical, and rabidly anti-Semitic. Hitler hated Jews. Katia, a Catholic, married a Jew. Kate, a Jew, married a Catholic. Eisner, the Jew, was prime minister of Catholic Bavaria. Jews, Catholics! As I reflected on these strange bonds, my thoughts drifted slowly back to Pauli. What was his religious background? I had assumed he was a Jew, but later research showed me it wasn't that simple. Pauli's grandfather was an elder of the Prague Jewish community, but his son moved to Vienna, changed the family name from Pascheles to Pauli, and converted to Catholicism to advance his university career. He also hid his roots: Wolfgang didn't know until he was sixteen that his father was Jewish, so Pauli, both a Jew and a Catholic, was himself representative of the blurring of boundaries that was so common in 1920s Munich.

My musings on that May morning then turned from 1918 to 1932 and back again to my mother. A few years after her arrival in Munich, Katia met Angelo Segrè, an educated, upper-class Italian Jew, and moved with him to Florence. However, they were both fond of Munich, often going back, so my older brother's birthplace in May 1932 was there, not Florence. A week earlier Kate, my future mother-in-law, delivered a baby girl in Munich, an older sister of my wife's.

The old 1932 photographs show how happy Katia and Kate both were. The future seemed promising for them and their babies. Admittedly the economic depression, not unique to Germany, was causing hardships, but that wasn't enough to dampen the young mothers' optimism. They believed things would be getting better. Instead, they got worse very quickly. Neither Katia nor Kate nor Pauli foresaw how Hitler would come to exert his control over Germany. Even in early 1933, after Hitler had already become Germany's chancellor and as restrictive legislation was being authorized, Pauli dismissed the notion that the Führer would seize total control. It was *Quatsch* (nonsense), he said vigorously to a group of friends. "I have seen dictatorship in Russia . . . In Germany it just couldn't happen."

But Pauli never was particularly astute or far-seeing in political matters, and at times he was forced to suffer the consequences of this deficiency. Though he could have left Europe much earlier, his harrowing escape to the United States did not take place until July 1940, well after World War II had started.

Chapter 2

The Changing Times

MEPHISTOPHELES
And yet for us dost work alone,
While thou for dam and bulwark carest;
Since thus for Neptune thou preparest,
The water-fiend, a mighty fête;
Before thee naught but ruin lies;
The elements are our allies;
Onward destruction strides elate,

Goethe, *Faust, Part II,* act 5, 503–9

The 1920s

IN MANY WAYS Hitler's takeover brought down the curtain on a period of extraordinary innovation in post–World War I Europe. The years between 1918 and 1933 were characterized by striking upheavals in art, social mores, thought, politics, and science and were probably the twentieth century's most dynamic period. It was an era of great optimism and wild experimenting, marked by James Joyce's cryptic retelling of the story of Ulysses, Arnold Schoenberg's atonal compositions, Giorgio de Chirico's eerie landscapes, Le Corbusier's manifesto for a new architecture, and Heisenberg's perplexing uncertainty principle.

Overarching themes, all centered on upending familiar tenets, began to appear. The outer world, carefully described for hundreds of years, did not seem to match the inner world that was being uncovered. Discussions of the id, the ego, and the superego, the new language of psychoanalysis, were appearing everywhere in authors' writings. The rules of pictorial representation, challenged by the Impressionists, were being jettisoned. The submicroscopic world of the atom did not seem to obey the known laws of physics.

Hierarchies and institutions that had preserved order before the Great War were crumbling, replaced by what some saw as emancipation and others as anarchy. Rebellious movements sprang up: surrealism proclaimed loosening of strictures through the liberation of the unconscious, and Dada, its very name a baby-talk expression, said underlying reality would be found through chance and irrationality.

The world's capitals were bubbling with excitement, with ideas, with fresh techniques. Admittedly many of the notions had begun to take shape earlier, perhaps in Paris or Vienna or even Milan. Filippo Tommaso Marinetti's 1909 "Futurist Manifesto" had declared that "museums are cemeteries" and "a racing car . . . is more beautiful than the Victory of Samothrace." However, the end of the war was the true watershed, the turning point in the new ideas' evolution. Marcel Duchamp had caused a furor at the 1913 New York Armory Show with his painting *Nude Descending a Staircase,* but after World War I he went a step further. Displaying copies of the *Mona Lisa* adorned with a mustache and a beard, he was throwing down the gauntlet to bourgeois gallery goers.

Centers of activity shifted among many of the big cities, national boundaries and local traditions no longer the obstacles they had been earlier. New Yorkers went to Paris, Parisians to Los Angeles, Londoners to Berlin, and Russians went wherever they could. Even Germany, crippled by the war, the political instability, and the inflation that followed, had its vital centers: Munich was one and Berlin certainly another. Bertolt Brecht and Kurt Weill may have caricatured the capital in their *Rise and Fall of the City of Mahagonny,* but the Berlin stage was at least as inventive as any in the world. George Grosz's finely etched drawings of rich and decadent burghers created a new form of social criticism, Fritz

Lang's futuristic movies portrayed unnatural criminal behavior, and Marlene Dietrich's outfits were both sexy and strangely androgynous.

Nor was art appreciation confined anymore to the so-called educated classes. The new media of photography and the cinema were accessible to all, though some asked if they were education or simply entertainment. Was psychoanalysis science, medicine, philosophy, or nonsense? Were Picasso's canvases great art or distortions? Were Mies van der Rohe's buildings classic or offensive? What about the old ideas of causality in science?

Longing for the stability and order that had once prevailed, conservatives viewed these so-called advances with apprehension and disapproval. What had happened to reliable Victorian morality? The flappers' short skirts, the nudity in cabarets, and the new dances seemed to them both obscene and offensive. All these novelties were taken as symbolic of the decay in moral fiber, the bacchanalia presaging the end. The German philosopher Oswald Spengler's 1922 *The Decline of the West,* purportedly describing exactly what its title says, was widely read and appreciated, quickly becoming an international best-seller. He envisioned the increasing mechanization of labor, the advent of the assembly line, and the new technologies acting in concert to diminish the human spirit. To illustrate his thesis, Spengler invoked the myth of Prometheus, punished by the gods after having stolen fire from them, drawing by analogy to this parable a pessimistic outlook for the West's future. In Spengler's view the decline was already well under way.

Some came to see the adoption of technology that Spengler described as a threat to their way of life. One much-discussed example used to illustrate this anxiety—I quote an excerpt from it as the epigraph of this chapter—was taken from *Part II* of Goethe's *Faust.* The troubled hero has embarked on a vast reclamation project, draining swamps and regaining land from the sea. This comes, however, at the cost of ruining the landscape and driving the elderly couple, Philemon and Baucis, from their home. The endeavor is finally shown to be futile, with Mephistopheles describing to the audience its final outcome.

Concerns like this and worries that their already hard lives would become even harder were much on the mind of ordinary working people. They longed for stability and economic security. After World War I, with the Kaiser gone, Germans looked to their new government to provide them with assurances, but their whole country, not just Munich, was in turmoil. In January 1919, after a preliminary set of elections, the National Assembly withdrew to write a constitution. Fearful of the unrest in the capital's streets, they decided to retire from Berlin to conduct their work in a place where they would not feel threatened. Searching for a site, these Germans naturally turned to Goethe, the symbol for them of culture, order, and enlightenment.

With his image in mind, the assembly decamped to Weimar, a small German city about 150 miles southwest of Berlin, once the seat of the grand duchy of Saxe-Weimar. It had been Goethe's home for more than fifty years. He had gone there in his midtwenties, already famous after the success of his novel *The Sorrows of Young Werther*, and had stayed in Weimar until his death. Because of Goethe's presence the city became a gathering place for artists and intellectuals, a pilgrimage site for all those who were interested in the higher spheres of learning, an oasis of quiet contemplation. Thanks to this heritage, Weimar continued to attract artists and thinkers, a tradition that endured even after World War I. For instance the Bauhaus, that shaper of modern art and design, guided by the likes of Walter Gropius, Paul Klee, Wassily Kandinsky, and Lyonel Feininger, was first established in Weimar in 1919.

Once the German constitution had been drafted there, the city gained a new worldwide connotation: Weimar Republic became the name of the new German government. Unfortunately, despite many successes in bringing the nation back to prosperity, its institutions failed to take hold. By 1930 the Weimar Republic was floundering, and its official end came in 1933 with Hitler's ascension to power. One year before that, as Germany was celebrating the hundredth anniversary of Goethe's death, Weimar gained the dubious honor of being one of the first regional administrations officially in the hands of the Nazi Party.

Five years later it would gain the much more ominous distinction of having the concentration camp of Buchenwald built on its outskirts, the very antithesis of all that Goethe had stood for. But those horrors were still unimaginable in 1932.

In April of that year, while the Weimar Republic was failing, a small group of physicists, perhaps three dozen or so, gathered for a meeting. They assembled in a Copenhagen lecture hall that was as full of meaning for them as the Weimar theater's stage had been in the halcyon days of Goethe and Schiller. They were the pioneers of quantum mechanics, the most significant and far-reaching scientific revolution of the twentieth century, thanks to the profundity of its ideas and the importance of its applications. Baffling and mysterious when first encountered, quantum mechanics represented a departure from older concepts every bit as radical as modernism had been in the arts.

The meeting they gathered for in 1932, informal in structure, with no set agenda, had been a yearly event since 1929. From 1931 on, it also included, de rigueur, a skit written and produced by the youngest participants. During its course the young would lampoon their elders, many of whom were still no more than thirty years old. Of course Bohr, the reigning monarch, was mocked as well. In 1932, with Goethe and Weimar in mind, the physicists decided the skit's subject would be a parody of *Faust*. The great debate between the Lord and Mephistopheles for the possession of Faust's soul would be adapted to one between Bohr and Pauli, cast respectively as the Lord and Mephisto. Faust's beloved Margarete, also known as Gretchen, would be the neutrino, at that time a bone of contention between Bohr and Pauli.

That's how the "Copenhagen Faust" came about, but I am getting ahead of my story. First we must learn how the quantum revolution started and why, years later, our characters gravitated toward Copenhagen. Then we will meet our dramatis personae.

The Birth of the Quantum

ON THE AFTERNOON of Sunday, October 7, 1900, Heinrich Rubens and his wife paid a visit to the Plancks, their neighbors and friends. The journey from one house to the other was an easy one because both families lived in Grunewald (German for green woods), the attractive suburb of Berlin at the edge of a small pine forest a little west of the city. The neighborhood was a favorite of Berlin professors. The houses were spacious, with large dining-room tables and ample cultivated gardens; they also included a comfortable study where Herr Professor could retire to work, perhaps with occasional but not too frequent interruptions from the children.

The visit that afternoon was nothing out of the normal. In German universities at the time, a department would have one senior professor of experimental physics and one of theoretical physics, the division between the two forms of research having taken place only a few years earlier. The Berlin positions were occupied in 1900 by, respectively, Heinrich Rubens and Max Planck, so the two of them, often cast together to settle administrative and student matters, had much to discuss. That afternoon, however, as their wives chatted, Rubens's conversation with his colleague soon turned to some puzzling results he had just obtained in his laboratory while measuring electromagnetic radiation emitted from an object kept at a fixed temperature. This was a subject near and dear to Planck's heart so he listened with great interest. Then, as is still true now, intense physics discussions between theorists and experimentalists were routine occurrences.

The experimental data, displayed as a plot of radiation intensity versus radiation wavelength, had agreed quite well up to then with a formula proposed years earlier by Wilhelm Wien, but Rubens had now extended the measurements to previously unobserved long wavelengths. He found that Wien's formula did not fit the new data. That evening, after Rubens left, Planck retired to his study to think about what the experimentalist had told him.

Theoretical physics often proceeds this way. Sometimes a new theory is the first step and data is sought afterward to prove or disprove it, but often the data itself is the starting point. One searches for a simple formula that matches the experimental results, and if it is found, the hunt is on to discover how the formula might be derived from known or new theories. The searcher needs to be experienced in data analysis and must know where innovation is possible within the established scientific framework, but intuition, experience, and a little luck all come into play at that point.

Decades may pass between obtaining the formula and explaining it, as was the case with Kepler's laws for planetary orbits. They had to await Isaac Newton for a true derivation. Fortunately in this case it took only a few months of intensive labor by Planck. His explanation was a radical one, causing him much anguish when he announced it. In his own words,

> It was an act of desperation . . . I knew the problem was fundamental and I knew the answer. I had to find a theoretical explanation at any cost, except for the inviolability of the first two laws of thermodynamics.

By saying he *knew the answer* he was acknowledging that he had a formula that fit the data right away on that first evening. Deriving it from physics principles was the hard part.

Planck had discovered that he needed to assume heated objects emit and absorb radiation in discrete packets of energy rather than in a continuous stream, as had always been assumed. He called the packets *quanta.* One could begin to understand their existence by imagining radiation subdivided into quanta as analogous to a jet of water composed of individual droplets. However, very surprisingly, Planck also found that his formula required energy emission and absorption to take place in multiples of a fundamental lowest-energy quantum. It was as if there were a smallest drop and all other drops' masses had to be multiples of the smallest one's. In addition, the ratio between the energy of the radiation quantum and its frequency was constant. This newly discovered constant, soon called *Planck's constant,* was the harbinger of quantum theory.

A little less than five years later, in March 1905, a twenty-six-year-old employee of the Swiss patent office took the next major step. Einstein daringly proposed that quanta move freely back and forth through space, absorbed or reflected by any obstacles they might encounter. In doing so, he was suggesting a particle-like nature for all electromagnetic radiation including visible light, the radiation with wavelengths in the range to which the human eye is sensitive.

Thinking of radiation as particles made it easier in some ways to understand the quantum picture. On the other hand, Newton's corpuscular theory of light had supposedly been disproved a century before Einstein's suggestion and replaced by a wave picture. This was the foundation of scientists' understanding of optics, and doubting it meant that a fundamental notion was being called into question. In Einstein's hands, quantum theory was creating new puzzles while explaining older ones.

Whereas Planck, because of his basic conservatism in physics, struggled mightily before accepting the concept he proposed, nobody would accuse young Albert Einstein of hesitating to overthrow traditional views. In the following years, as Einstein continued with increasing success to pursue his picture of the quantum, the doubts about its veracity slowly evaporated. But the dual nature of radiation, wave or particle, remained a conundrum.

Quantum theory's next major step came through attempting to understand how radiation could be emitted and absorbed by atoms. The existence of these units of matter, the smallest entities of an element, had been a subject of debate in the late nineteenth century, but by 1910 there was little doubt of atoms' reality, only a question of their makeup. Ernest Rutherford, arguably the twentieth century's greatest experimental physicist, put that topic to rest in 1911. According to him, and contrary to earlier views, the atom is mostly a great void, much like our own solar system. It holds in its core a comparatively tiny nucleus, a "fly in the cathedral" as it was sometimes referred to after Rutherford's discovery. Rutherford himself, using for comparison the great London auditorium rather than a more ecclesiastical setting, referred to the nucleus as "a gnat in Albert Hall."

The discovery of the atomic nucleus ranks as one of the most important scientific findings of the century, and a surprising one at that. We now consider the experiment that produced it the turning point in all explorations of the subatomic world and the prototype of how to explore this mysterious domain. The present-day billion-dollar superconducting colliders are its direct descendants. At the time, however, the impact of Rutherford's experiment was not immediate because physicists had very little understanding of the atom's overall structure, of how and why electrons move around the nucleus, and how atoms bind to one another to form molecules. Until some of these questions were settled, physicists thought it was premature to worry about the composition of the much smaller nucleus.

It took another physicist to see how to join Rutherford's picture of the atom to Planck and Einstein's picture of the quantum. The solution to that puzzle began to become clear in 1913 through the work of a twenty-seven-year-old Dane named Niels Bohr.

Why Copenhagen?

IN SEPTEMBER 1911, shortly after receiving his physics doctorate in Copenhagen, Niels Bohr went to England, funded by a Danish fellowship. A period at Cambridge University proved rather unproductive for the shy and awkward young man who spoke only halting English in a soft, low voice. But a few months after his arrival, he managed to transfer to the University of Manchester, where Rutherford's research was in full swing. This stay turned out to be decisive for Bohr, both personally and professionally. Bohr would later say of Rutherford, "To me, he had almost been like a second father." The close friendship between the older, already famous experimentalist and the young theorist lasted a quarter century; it ended sadly with Rutherford's sudden death from a botched hernia operation.

Rutherford's influence on Bohr went beyond just encouraging him to solve atomic physics problems. The young Dane saw in him a style of collaboration achieved by surrounding oneself with talented young people and then both guiding and inspiring their work. Rutherford's Manchester group came from around the world and from all social classes. Charles Darwin, grandson of the biologist, was from an upper-class Cambridge family, while James Chadwick's roots were working-class Manchester. Georg von Hevesy was a Jewish Hungarian nobleman, while Hans Geiger was German and later discovered to be anti-Semitic. Ernest Marsden had a fellowship from Rutherford's native New Zealand, and Bohr one from Denmark. After they left his institute, Rutherford followed his young protégés' careers and helped when he could. Some called this assistance paternalistic, but the motivation was always to push the boundaries of what one knew in physics and to help "his boys."

Rutherford was extraordinarily ingenious, and the Manchester experiments tended to be quite simple, with most of the equipment built and assembled by the experimentalists themselves at relatively little cost. He had made his name by studying radioactivity—his Nobel Prize was in chemistry, not physics—and the trademark Rutherford experiment used a beam of particles emitted by radioactive substances. The beam, aimed at a thin target, was scattered as it passed through the target, and the results of that scattering were then analyzed. His laboratory discovered the atomic nucleus this way, and twenty years later, using the same approach, it would be the first to successfully produce a nuclear disintegration.

As Rutherford grew older and became more of a leader than a hands-on experimentalist, his formidable intuition, enormous energy and enthusiasm, and great dedication to his work continued to provide an example to the younger physicists surrounding him. He would walk around his laboratory supervising work, helping "his boys," giving support and encouragement. Though his knowledge of theoretical physics was slight and he was suspicious of complicated mathematical expressions, he nevertheless believed in having theorists guide his experimental work and often sought their advice.

Manchester's collective atmosphere was extraordinarily inspiring for young Bohr. Even though he was not an experimentalist, he loved the daily discussions, the sense of relying on one's intuition. Above all, he admired the constant reminder that one should always aim to solve important problems.

After a short but crucial stay in Manchester, Niels Bohr returned to Copenhagen. Trying to launch his academic career, he found himself stymied by the absence of an appropriate position in his native country. So, in 1914, Bohr returned to Manchester with an appointment equivalent to an associate professorship. In the meantime he had married the beautiful and charming Margrethe, whom he had met before his initial journey to England. Their 1912 honeymoon trip had included a visit to Manchester, during which a long friendship between the Bohrs and the Rutherfords was cemented.

Denmark, now concerned that it might lose Bohr, responded by creating its first professorship in theoretical physics and appointing him to fill the position. Rutherford tried to counter the offer and keep him in Manchester, holding forth the prospect of the great work they would do together. But the pull of home was too strong for the Bohrs, and home was always Denmark. Niels and Margrethe returned to Copenhagen in 1916.

At first Bohr's office space was a modest ten-by-fifteen-foot room that he shared with a young Dutchman, Hendrik Kramers, the first of many foreigners who came to Copenhagen to work on quantum theory with the new professor. When in 1919 Bohr hired a secretary, her desk was placed in that same small room.

But Bohr immediately began planning something bigger. Tirelessly working to raise funds from foundations, the government, and private sources, he soon had enough money for a building on a shady Copenhagen street, the Blegdamsvej. Constructed in a neoclassical style, it had four floors, of which the first was partially below ground level. Stepping up from the street to its big double doors, above which was inscribed "Universitets Institut for Teoretisk Fysik," visitors entered a large hallway with a lecture hall containing graded seats located to the right. The building housed a modest library, some small offices, and a lunchroom where one could always find coffee and good Danish bread with cheese.

Upstairs there was an apartment for the growing Bohr family, now numbering three sons: Christian born in 1916, Hans in 1918, and Erik in 1920. Eventually there were three more sons: Aage born in 1922, Ernest (named after Rutherford) in 1924, and Harald in 1928. These boys would in later years have as baby-sitters and companions in soccer games and sailing trips some of the world's most famous theoretical physicists, several of them not much older than Christian.

The Institut for Teoretisk Fysik, or more simply Bohr's institute, opened in 1921. By 1922 operations were in full swing, and Bohr was ready to once again direct his full attention to physics. It was an auspicious year for him. As a confirmation of his rising status in the physics world, Bohr, now thirty-seven, was awarded the 1922 Nobel Prize in Physics.

A few months earlier, in June, he had gone to Göttingen to give a series of seven lectures on quantum theory. Since this small German town's university had one of the world's finest physics and mathematics faculties, the announcement of what subsequently came to be known as the Bohr Festspiele (Bohr Festival) was widely circulated. The series attracted young physicists from all over northern Europe. Many, having never met Bohr before, were enchanted by him; their changed attitude toward physics reflected this. As Heisenberg would say later, "I learned optimism from Sommerfeld, mathematics in Göttingen and physics from Bohr." At the time, the German approach to theoretical physics was formulaic: commence with an equation, solve it, display its solutions as generally as possible, and then begin to analyze the consequences. Bohr started with a picture, then maybe a thought, and perhaps a dimly viewed connection. He then stitched them together; the equation came last. It was only the formalization of what he had already glimpsed.

The combination of this new approach and Bohr's personality was a powerful magnet. Although recruiting had not been the intent of Bohr's visit, theoretical physicists began flocking to Copenhagen and its fame spread. During the 1920s more than sixty theorists visited Bohr's institute for prolonged stays, many remaining for years. They came from around the world, as far away as the United States, Russia, and Japan. Some came with fellowships from their home countries while others were supported through funds raised by Bohr. Almost all of them were young; they lived together, ate together, played together, and worked together. Their manner marked a new informality in scientific circles, with no distinction between assistant and Herr Professor. There was equal treatment for ideas originating from the youngest or the oldest, the fresh PhD or the Nobel Prize winner. Bohr went walking, skiing, and even to the movies with the young physicists who surrounded him, and they played soccer with his sons. But the overarching context, both at play and at work, was quantum physics.

The Meetings Begin

BOHR'S WARMTH, enthusiasm, energy, and intelligence weren't the only reasons for his institute's success. He was also lucky. Puzzling out the atom's workings turned out to be the central physics problem of the 1920s. Furthermore the subject's development in that period was unusual in that theoretical advances were far more important than experimental ones. This is not to say that important experiments were not performed during the decade, but rather the key progress was due to bold new ideas.

Bohr's explanation of the hydrogen atom, proposed originally in

1913, and its subsequent refinements explained a great deal of experimental data so well that by the early 1920s physicists felt there had to be some truth in it. However, by 1925 the model's failures in providing satisfactory explanations of the stability of atoms, the details of the periodic table of elements, the absorption and emission of radiation by atoms, and the complex behavior of atoms in external electric and magnetic fields all pointed to the need for serious modifications in understanding how electrons behave inside an atom.

Bohr's young disciples Pauli and Heisenberg were probably the two who felt most strongly that a radical break with quantum theory's short past was needed. It wasn't enough to make changes here and there, while keeping the basic framework. In 1925 Heisenberg went even a step further, stating that the heretofore-observed agreement between theory and experiment was accidental: electron orbits were meaningless. As he says on page 1 of his groundbreaking 1925 paper, the paper that gave the first hint of the new quantum mechanics,

> In this situation it seems sensible to discard all hope of observing hitherto unobservable quantities, such as the position and period of the electron, and to concede that the partial agreement of the quantum rules with experience is more or less fortuitous.

A *more or less fortuitous* agreement certainly wasn't good enough for Heisenberg. A new theory was needed, and he was hoping to provide the first step along that path. What it meant to measure an electron's position and velocity had to be understood. Long and hard thinking, new insights into what happens in the mysterious subatomic world, were necessary, not just more experiments.

In this situation the existence of a place where the theoretical physicists could meet and exchange ideas with one another without feeling confined by hierarchical structures, where all would be judged solely on the strength of their ideas, was both novel and invaluable. Bohr's institute was that place, and its contribution to the development of quantum physics during the 1920s was crucial. This was never more true than during 1926 and 1927, when Bohr and Heisenberg worked out what they and Pauli came to call the Copenhagen interpretation of quantum mechanics, the still-accepted view of the subject.

By 1929, Bohr was at the height of his powers and fame, his institute a legend. In the spring of that year he received letters from two of his favorite theoretical physicists, both saying they were hoping to take advantage of the weeklong Easter university vacation to visit him. The first letter was from Hendrik Kramers, who had returned to the Netherlands and taken up a professorship in Utrecht after a decade in Copenhagen. The second came from Wolfgang Pauli, by now generally recognized as having the finest critical mind in quantum theory. Bohr was thrilled to welcome them back. Then he had an idea of how to make the reunion even better. Why not invite twenty or more of the former Copenhageners for a free-ranging week of discussions? No agenda, no set topics, no formality, just give-and-take about whatever seemed interesting in quantum physics. It would recapture on a larger scale the institute's spirit of free discussions, open criticism, and wide collaboration.

Invitations went out. Heisenberg had left in March for a trip around the world, but many others would be able to come. The majority of them were still very young, but one of them, older than Bohr, came to mind immediately. He was Bohr's very good friend Paul Ehrenfest, a Viennese-born theorist and now a professor in the Netherlands, at Leiden. Ehrenfest was a spectacular teacher and a famed clear thinker. His original work might not have been on the level of Bohr's or Einstein's, but then whose work was?

Bohr also encouraged invitees to bring along particularly bright students. Ehrenfest took him up on the offer and arrived in Copenhagen with twenty-year-old Hendrik Casimir. In his memoirs Casimir remembers the leisurely journey from Holland; they left Leiden early Saturday morning for the quick ride to Amsterdam. Then came a daylong train trip to Hamburg; after a night in a small Hamburg hotel they were back en route. As their train approached Warnemünde, where they caught the old ferry across the Baltic to Denmark, Ehrenfest suddenly fell silent. Turning to Casimir, he said, "Now you are going to know Niels Bohr and that is the most important thing to happen in the life of a young physicist," and then he smiled.

Ehrenfest saw old friends on the boat. Chatting away, they were all soon on yet another train for the short ride to Copenhagen, to be greeted at the station by Bohr and his sons. A festive dinner came next, and then it was time for bed. Casimir was asked to share a boardinghouse room with a brilliant twenty-four-year-old Russian physicist named George Gamow. He tried to get some sleep but was too excited.

The next morning Ehrenfest formally introduced Casimir to Bohr. With a hand on his student's shoulder, Ehrenfest said to his Danish friend, "I am bringing you this boy. He has already some ability, but he still needs thrashing." Impressing Bohr favorably during the following days, Casimir was invited to remain in Copenhagen after the meeting, a stay that eventually lasted almost two years. His parents, concerned about the sudden separation from their young son, weren't altogether reassured by the glowing terms in which their Hendrik portrayed the Copenhagen atmosphere. However, they were put at ease once a letter simply addressed to Hendrik Casimir c/o Niels Bohr, Denmark, reached Hendrik without delay. They now realized that their son's protector was no ordinary scientist or citizen.

The tradition set in 1929 of the annual weeklong Copenhagen meeting over the Easter holiday lasted until the onset of World War II. The meeting itself became synonomous with a new style, led by the likes of Bohr, Pauli, and Heisenberg, of free exchange without regard to rank or age. Meanwhile, as so often happens with successful spontaneous acts, its reach expanded as the years went by. Though the essential informal nature of the gathering remained intact, after 1932 an effort was made to include those who were not old Copenhageners, as long as they had something to contribute.

That's how my uncle Emilio happened to go to the 1937 meeting.

Emilio Segrè was lucky in his timing. When still a twenty-two-year-old engineering student at the University of Rome, he was told of the arrival in the Italian capital of a new professor of physics, supposedly a genius and, at twenty-six, unimaginably young by the standards of the tradition-bound faculty in Rome. That genius, Enrico Fermi, slightly older than Heisenberg and a year younger than Pauli, was looking for bright young engineers who might want to work with him, as there were no physics graduate students in Rome at the time. Possessing a rare combination of both theoretical and experimental gifts, Fermi went on to become one of the greats of twentieth-century science, but back then he was just a beginning new professor looking for students.

My uncle became his first, commencing a collaboration with Fermi that was intense for the next decade, intermittent as they went their separate ways, and intense again during the two years they spent together at Los Alamos, New Mexico. Emilio had a distinguished career in his own right, including a Nobel Prize in Physics awarded in 1959, but he always remembered how much of his success he owed to Fermi.

In 1937 Emilio was still a relative newcomer, but having done some interesting research the year before, he received an invitation to go to Copenhagen, one he was honored to accept. The number of attendees had increased since the very early days, but other features of the meeting were much like earlier years': Bohr sat in the lecture hall's front row, with Heisenberg and Pauli next to him. He still interrupted talks in his polite but persistent way while Pauli was, as usual, equally persistent but more caustic. But the general mood that year, with a pall cast by the pervading political climate, was not carefree as it had previously been. Right after the meeting my uncle wrote to a cousin in Rome, setting down his impressions of Bohr, the 1937 skit, and hints of what was happening in Germany:

> Yesterday evening the Congress ended, with a humorous, but rather moving, feast. We acted in a sort of variety show summarizing

> Bohr's recent travels around the world. Through the jokes one could feel the respect and almost veneration that everybody feels for Bohr. I could not approach him very much, but I understood that he is one of the most remarkable personalities produced by mankind, and that he hovers in heights incomparably higher than those reached by common mortals, be they even Fermis. Also morally and from a human point he must be superior to others. Immediately after the feast I left with Heisenberg and his wife. Heisenberg . . . has been a pupil of Bohr's at Copenhagen for three years and did his best work there. Bohr said a few words of good-bye to him and his wife that well nigh made the company shiver and everybody was clearly shaken.

In 1937 Hitler's regime was in full ascendancy. Heisenberg, seeking an accommodation with the Nazi regime, was coming under attack in Germany for not being anti-Jewish enough. Outside of Germany he was viewed as a willing collaborator of the system and therefore tacitly anti-Jewish. My uncle, an Italian Jew and married to a German Jew, was certainly sensitive to those nuances, as was Bohr, half-Jewish and by now the leader of the movement to try to protect refugee scientists. Bohr's institute had served for many years as a center of scientific exchange; it now had become the haven for fleeing physicists, the first stop in their attempts to relocate. We can see their shining faces in a 1937 conference group picture: Maurice Goldhaber, Lise Meitner, Rudolf Peierls, George Placzek, Emilio Segrè, Otto Stern, Victor Weisskopf. Many had already left Italy or Germany by then and the rest would soon follow.

But five years earlier, there had been no such turmoil.

The 1932 Meeting

By 1932 attendance at the Copenhagen meeting had already risen to nearly forty, almost filling the institute's lecture hall. Though Pauli had come to the first three conferences, in 1929, 1930, and 1931, he was not able to attend in 1932, as Heisenberg had not in 1929. Bohr understood that former participants could not always come back, but a core of them would be enough to ensure lively weeklong discussions.

There are several pictures of the 1932 gathering, but my favorite one of this story's personages, though minus Pauli, was actually taken at the 1933 Copenhagen meeting (see photograph 14). The first of the six, looking from left to right in the front row, is Bohr. Next to him, in the position that Pauli might have occupied had he not missed the meeting again, is Paul Dirac, a shy and ascetic British genius. Continuing along the row, there is Werner Heisenberg, now thirty one years old but looking younger, and then Paul Ehrenfest, who once again had come from the Netherlands. Next to him, eyes downward, averted from the camera, is twenty-six-year-old Max Delbrück, probably feeling awkward at being in the front row of such an illustrious gathering. Perhaps he is there because he was a favorite of Bohr's. The last person in the row is Lise Meitner, a good friend of Bohr's for over a decade, the only woman in the front row and the only experimentalist.

The grouping of seven is interesting for several reasons. First is the sheer brainpower. Bohr, Pauli, Heisenberg, and Dirac are titans in the world of physics. Bohr is the father figure, the guide and unifier, Pauli the most critical, Heisenberg the most intellectually daring, and Dirac the most original. Ehrenfest is not quite in that league, but his ability to quickly grasp the key features of a problem made him an invaluable contributor to the gathering.

Delbrück, the youngster in the crowd, would slowly switch, with Bohr's encouragement, from physics to biology, eventually becoming an iconic figure in molecular biology, a Nobel Prize winner as well, but in physiology or medicine, not physics. He was also the author of the 1932 skit, the "Copenhagen Faust," sometimes known as the "Blegdamsvej Faust" after the street on which Bohr's institute was located.

Then there is Lise Meitner. She was an experimentalist among theorists at the meeting, though far from the only experimentalist to ever be invited, for Bohr always recognized how critical it was for the two kinds of physicists to keep each other informed about what they were doing and what they were thinking of doing. Having learned the importance of these exchanges early on during his stay with Rutherford in Manchester, he made sure they continued to be part of his life. A small but significant part of his institute was always devoted to experimental research, just as Rutherford had devoted a small but significant part of his to theory research.

Of course some experimentalists were better than others at explaining what they were doing, at highlighting possible measurement errors, and at guiding theorists to fruitful research. Meitner was one of the best. She was also a very accomplished practitioner in her own right, performing over the course of half a century a large number of important experiments.

However, she is best known for a sudden flash of insight that came to her while strolling with her nephew Otto Frisch in the Swedish woods near the town of Kungälv on the morning of December 24, 1938. That day in 1938 the two of them were the first to understand that when struck by a single neutron, a uranium nucleus can split into two pieces releasing a considerable amount of energy. Frisch soon gave this process, the key to the atom bomb, the name *fission*. Meitner probably deserved the Nobel Prize on her own or with her longtime collaborator Otto Hahn for her cumulative experimental work; she certainly deserved it with Frisch for this insight. Bohr recommended that it be awarded to her, but it wasn't. The Meitner case is now regarded as an example of wrongheaded political interference that occurs even in the selection of the three science Nobels. It's tempting to think this was a case of bias against women scientists, but it probably had more to do with an unwarranted personal animosity toward her that some prominent Swedish scientists felt. But her contributions were real, prize or no prize.

The seven are also interesting because they represent three generations of quantum theory physicists. Bohr, Ehrenfest, and Meitner were members of the older generation, contemporaries of Einstein. All three received their education and were working physicists before the quantum mechanics revolution of the mid-1920s. Pauli, Heisenberg, and

Dirac, coming of age with the revolution, were key players in the shaping of the new theory. Finally Delbrück, though only six years younger than Pauli and four years younger than Dirac, belonged to the generation "after the revolution," those who began their careers when the main features of quantum mechanics were already in place.

There is, however, another presence in Copenhagen during that week that cannot be ignored if we are going to tell the story of the "Copenhagen Faust." His name is Johann Wolfgang von Goethe.

Chapter 3

Goethe and *Faust*

FAUST *[DYING WORDS]*
Nor can the traces of my earthly day
Through ages from the world depart!
In the presentiment of such high bliss,
The highest moment I enjoy—'tis this.

Goethe, *Faust, Part II*, act 5, 542–45

In the Glow of Goethe

WHEN NAPOLEON, a longtime admirer, finally met Goethe, he is reported to have declared, "Voilà un homme," "Here is a man." Goethe was "the Wisest of our Time," as Thomas Carlyle, his English champion, called him. Admirers came from all over Europe to pay homage to the great one in the new Athens that rose around him in Weimar. Even after his death, they made the pilgrimage to his grave and to walk on the hallowed grounds on which he had trod. In the wake of World War I, many of the old idols were crumbling, but admiration of Goethe and respect for him remained; he still represented true eminence of mind and spirit.

Goethe's life has been documented in extraordinary detail, starting with his own renditions. He recorded every aspect of that life, or as he said it in his own famous aphorism, all his writings were "fragments of a great confession." Those so-called fragments amount to more than a hundred volumes in the standard edition of his works, which include travel writing, speeches, government reports, and scientific works as well as poems, novels, and dramas.

Fame came at age twenty-five to Goethe with the 1774 publication of *The Sorrows of Young Werther.* Narrating the tragic story of Werther's unrequited love of Charlotte, betrothed to Albert, this short novel quickly became a sensation throughout Europe. Young men emulated Werther in speech and dress, occasionally even in committing suicide, mimicking the story's heartbreaking ending. Though the novel was loosely based on an unhappy love of Goethe's, its author, unlike Werther, recovered quickly and went on to have many other loves, some unrequited and others reciprocated.

Shortly after this first novel's triumph, Goethe moved to Weimar. He went there at the invitation of Karl Augustus, duke of Saxe-Weimar, the eighteen-year-old absolute ruler of this small independent state in eastern Germany. Though Weimar, with fewer than ten thousand inhabitants, had already become somewhat known as a cultural center thanks to the influence of the duke's mother, it was a far cry from the great commercial center of Frankfurt, where Goethe had grown up and where he was living when the invitation reached him. But the opportunity to move there struck a chord in the young writer, and he quickly accepted the duke's offer. He almost certainly made the right choice, since Weimar provided Goethe with an ideal setting to pursue his many interests and, significantly, with financial security. In turn Goethe repaid the duke's generosity by making Weimar synonymous with learning and culture throughout the world.

During the first decade of his Weimar stay, Goethe participated in the city's administration. Trained as a lawyer and having spent some time practicing that profession, Goethe was well versed in legal matters. He had also by then begun his scientific investigations, studied some medicine, learned the craft of etching, and of course written a great deal. *Faust*'s famous opening monologue echoes such a list:

I have, alas! Philosophy,
Medicine, Jurisprudence too,
And to my cost Theology,
With ardent labor, studied through.
And here I stand, with all my lore,
Poor fool, no wiser than before.

(*Faust, Part I,* 1–6)

This speech, so easily modified, has often been parodied, including in the "Copenhagen Faust," where a recital of various subfields of theoretical physics makes up the catalog of unsatisfying attempts at knowledge. However, Goethe, far from being a "poor fool," flourished in Weimar. Having been appointed privy legation councillor, Goethe took an active interest in many of the duchy's affairs and was involved in organizing the treasury, building roads, landscaping the parks, and reopening the district's silver, lead, and copper mines. Of course he continued his own writing as well.

After his first decade of intense activity in Weimar, Goethe felt the need to escape its confines. With the duke's blessing, and a stipend from him that encountered some criticism in Weimar because of its generosity, Goethe went south, spending most of the next two years in Italy, principally in Rome. There he dedicated a great deal of time to drawing, an activity that had always fascinated him, but he ultimately became convinced that his true talent lay in the written word.

Refreshed and now ready to channel his energies to the pursuit of his muse and to the cultural life of Weimar, Goethe returned there in 1788. For the next forty plus years, except for brief trips, he remained in this small city in Saxony. With his patron's blessing, Goethe was, on his return, freed of administrative duties other than those that intrigued him.

One particularly dear interest for Goethe was the theater. The decade between 1794 and 1805, when Friedrich von Schiller, the other great German dramatist of the period, joined him in Weimar, was a magnificent era for the German stage, marked by the duo's writing as well as by their mounting and staging of productions. Schiller's finest work, the idealistic historical dramas *Mary Stuart*, *William Tell*, and *Wallenstein*, all date from these years, as does much of Goethe's work on *Faust, Part I.* Sadly, Schiller's early death in 1805 put an end to their partnership. Goethe died twenty-seven years later. His head crowned with laurels, Goethe's body was placed in what is known as the Princes' Tomb, where his coffin lies side by side with Schiller's, a memorial to the great days of both the German stage and of Weimar.

The reverence expressed for Goethe was undoubtedly shared by the physicists in Copenhagen. There was, however, another link for them to the great humanist, that of science. Even before his work on the mines in Weimar, Goethe had been fascinated by geology, in particular by the subject of the earth's crust formation. This complemented his views on biology, particularly the existence of *Urpflanze*, archetypal plants from which others gradually took form. Nor was Goethe's curiosity limited to plants and rocks. A great deal of credit was given to him for his discovery of the intermaxillary bone's existence in the human jaw. At the time most anatomists maintained this structure was only present in animals, whereas Goethe saw correctly that it was simply hard to recognize in humans. His holistic belief of animal, vegetable, and mineral life lying on a continuum has even led many to make the stretch of calling Goethe a precursor of Darwin.

At the 1967 meeting in Weimar of the Goethe Society, Heisenberg characterized the master's views on science this way: "For Goethe all observation and understanding of nature began with the immediate sensory impression; not therefore with an isolated phenomenon, filtered out with instruments and so to speak wrung from nature, but with the free natural happening, directly accessible to our senses." Goethe expresses that approach by having Faust, after he has lamented the uselessness of his own book learning, urge such a communion with nature:

Up! Forth into the distant land!

. . . .

The courses of the stars unroll'd;
When nature doth her thoughts unfold
To thee, thy soul shall rise, and seek
Communion high with her to hold,
As spirit doth with spirit speak!

(*Faust, Part I,* 65, 69–73)

Goethe's enthusiasm was, however, overlaid with a disdain, inspired by Romanticism, for the kind of precise quantitative analysis that usually accompanies successful scientific inspiration. This was not an accident; as Mephistopheles says in *Faust, Part II,*

Herein your learned men I recognize!
What you touch not, miles distant from you lies;
What you grasp not, is naught in sooth to you;
What you count not, cannot, you deem, be true;
What you weigh not, that hath for you no weight;
What you coin not, you're sure is counterfeit.

(*Faust, Part II,* act 1, 306–11)

Goethe's *Theory of Colors,* the culmination of twenty years of investigations, remains his best-known scientific work. Though it first appeared in 1810, he had worked on it ever since his trip to Italy twenty years earlier. In this treatise, Goethe tries to incorporate notions of light and dark, cold and warm, harmony, repulsion, and comfort in a sort of aesthetic of colors. Couched in quasi-scientific terms with yellow and blue as the only totally pure colors, Goethe meant his work as a counterweight to Newton's unfeeling but empirical theory of colors; Newton's theory withstood the challenge.

Goethe, whom many call the last person to encompass all branches of knowledge, believed his contributions to science were almost on a par with those he made to literature. Indeed, his research in both physics and biology is thought-provoking, certainly not to be dismissed. Furthermore, his efforts in these directions and more broadly in shaping an all-encompassing approach to human emotions, science and poetry, truth and beauty, help to explain his depiction of Faust. They were also deeply influential for many scientists, even if their approach to science was very different from his. Goethe also reminded them that, in Mephistopheles' words,

> **Gray is, young friend, all theory:**
> **And green of life the golden tree.**
>
> (*Faust, Part I,* 1684–85)

The "Copenhagen Faust"

THE IDEA OF putting on a skit during the Copenhagen conference apparently originated during early 1931 in the mind of the wildly imaginative Russian physicist George Gamow, the same man who shared a room with Casimir on the latter's first night in Copenhagen. He envisioned it as a parody of a contemporary spy movie called *The Stolen Bacteria* that many of the physicists had seen at the local movie theater. Performed at the conference, the skit was a big success, thus initiating a new add-on to the serious physics discussions that marked the yearly gathering. Gamow always loved games, parties, puns, and elaborate practical jokes. His humor was later put to widespread use in more than a dozen popular books, all of them illustrated by the author. Translated into many languages, these books were often the first encounter for budding scientists of my generation with the world of modern science, ushering us into it with wit, charm, and a formidable bank of knowledge behind Gamow's easily approachable style.

Under ordinary circumstances Gamow would have also written the 1932 skit or at least helped Delbrück with its production. The two of them had been close friends in Copenhagen and had even written a paper on the subject of nuclear physics the year before. But these were not ordinary times. After lengthy stays in Copenhagen and Cambridge, Gamow had returned to the Soviet Union in late 1931. The Soviet government, clamping down on foreign travel, denied his request for a passport renewal, making a trip to Copenhagen in 1932 impossible.

Delbrück, reluctant to leave his absent comrade entirely out of the picture, added a short interlude to the skit, intending it as an homage to his missing friend. In the beginning of one scene, toward the back of the stage, a strange painting appears. It is of Gamow, hands clutching the bars of a jail cell. A voice from behind the stage is heard intoning sadly,

I cannot go to Blegdamsvej
(Potential barrier too high!).

This double entendre was an inside joke, typical of the skit as a whole. The key to nuclear disintegration, very much on physicists' minds in 1932, is the ability of a particle inside the nucleus to pass to the outside.

Gamow behind bars.

This probability, first calculated by Gamow in 1928, depends crucially on the barrier created by forces within the nucleus. These effectively keep the nucleus's contents within its confines most of the time for some nuclei and all of the time for others. But different barriers were on Gamow's mind in 1932.

Physicists were just beginning to learn that getting out of the Soviet Union was a formidable task, so in this case the "potential barrier" also had a clear political significance: Gamow was trapped in his homeland against his will. The obstacles he faced had apparently become too great even for someone as resourceful as he was.

After failing to obtain a passport, Gamow tried other ways of getting out of the country. Together with his first wife, he attempted escape in 1932 by rowing across the Black Sea to Turkey in a kayak of sorts. The only document Gamow carried with him was an expired Danish driver's license. On reaching Turkish soil, he intended to pass himself off as a Dane and ask to be taken to the Danish embassy in Istanbul. From there he thought his improbable odyssey would be over because, as Gamow wrote, "I would telephone Niels Bohr in Copenhagen, and he would arrange matters."

The story's interest is in its illustration of the confidence young theoretical physicists at the time had in Bohr: not only would he do anything to help them, but he also had the power to fix anything. The Gamows' attempted escape failed, rough seas pushing them back to Crimea. Luckily, through an extraordinary set of fortuitous accidents and now with direct help from Bohr, they managed a few months later to obtain passports to attend a Western European physics conference. By late 1933 they were out of Russia for good.

If not for Gamow and his sense of humor, the script of the 1932 "Copenhagen Faust" might have survived only in the memory of the participants. Though he couldn't attend the meeting, Gamow is responsible for seeing that it appeared in print, albeit in English rather than the original German. Two years before his premature death in 1968, he published a small book entitled *Thirty Years That Shook Physics: The Story of Quantum Theory*. It concludes with a translated version of the skit and a slightly cryptic comment from Gamow:

> The Blegdamsvej *Faust*, rendered into English by Barbara Gamow, is reproduced in this book as an important document pertaining to these turbulent years in the development of physics. The authors and the performers prefer to remain anonymous, except for J. W. von Goethe . . .
>
> Thanks are due to Professor Max Delbrück for his kind help in the interpretation of certain parts of the play.

Gamow had in fact obtained a copy of the script from Delbrück, translated it with the aid of Barbara, his second wife, and then charmingly illustrated it.

All we really know about the script's preparation is that Delbrück was its principal author, probably aided by Carl Friedrich von Weizsäcker, a twenty-year-old protégé of Heisenberg's. It seems safe to guess that though the 1931 skit had been a success, the physicists wanted something grander in 1932. It was, after all, the tenth anniversary of Bohr's institute.

Delbrück's choice was a natural one since celebrations were planned all over the German-speaking world for the one hundredth anniversary of Goethe's death. If the likes of Bohr were titans, Goethe was a god, the last person to encompass within his scope all branches of knowledge, and *Faust* was his masterpiece. In the 1920s students still learned by heart large sections of poetry. All German high school students had read the epic poem, and many of them had memorized lines from the drama of the conflict between the Lord and Mephistopheles for the possession of Faust's soul. Moreover even those who had forgotten the drama's verses were familiar with the plot and would be able to appreciate the nuances of a takeoff.

Bohr's institute in 1932.

Nor was the admiration for Goethe limited to the German-speaking world. For instance, one of Bohr's biographers, describing life in the Danish household while Niels was growing up, wrote that

> Professor Bohr [Niels's father], in addition to his lifelong absorption in physiology and natural science, was a disciple of Goethe and could recite whole sections of *Faust* from memory. Niels, as he walked at his father's side or sat at his feet on long winter evenings, learned Goethe almost by absorption. The majestic lines stayed with him all his life and he frequently quoted from Goethe.

This was still the time when members of the "educated class" were expected to be thoroughly familiar with certain works, and *Faust* ranked among the greats in the canon of Western literature. James Joyce's *Finnegans Wake* lists "Daunty, Gouty and Shopkeeper" as the three universally acclaimed poets, they being of course Dante, Goethe, and Shakespeare. *Faust, Part I* was particularly well-known because the story had been used and adapted so many times following Goethe's version, itself a variation of a sixteenth-century story powerfully staged by Christopher Marlowe in *The Tragicall History of the Life and Death of Doctor Faustus.* After Goethe, it was almost impossible to escape knowing of Faust's love for Margarete (also known as Marguerite, Margaret, Margareta, or Gretchen) and his pact with the devil.

Faust's popularity was further increased by the adoption of its subject in numerous other venues. Schubert's lieder based on some of the drama's poems were among the most loved of German song cycles. Of the dozens of operas with themes from *Faust,* Berlioz's *La Damnation de Faust,* Gounod's *Faust,* and Boito's *Mefistofele* quickly took their places in the standard repertoire. New York's Metropolitan Opera opened its doors in 1883 with a production of Gounod's version of the story and presented it so often in the first years that New Yorkers started referring to their new theater as the Faustspielhaus, a pun on the German word for festival hall. By 1932, cinema versions of *Faust* were also being shown, with F. W. Murnau's a particular favorite in Germany. In short, Europeans had to virtually isolate themselves completely from society and all art forms to escape knowing the plot of *Faust.*

Even beyond art, the metaphor of a Faustian bargain was consistently being used, and still is, as a shorthand for trade-offs in life that entail immediate gratification at the peril of greater loss in the long run, whether personal, political, or professional.

In those troubling political currents that characterized the early 1930s, physicists realized they might be called on to make some Faustian bargains in their personal lives, but they consoled themselves with the thought that their science was pure, abstract, with no dangerous unintended consequences. There would be no risk of Faustian bargains regarding their work. A decade later, as the full power of the atomic nucleus was unleashed, they would see how wrong they had been.

Chapter 4

The Front Row: The Old Guard

MEPHISTOPHELES *(ALONE)*
The ancient one I like sometimes to see,
And not to break with him am always civil;
'Tis courteous in so great a lord as he,
To speak so kindly even to the devil.

Goethe, *Faust,* Prologue, 108–11

Niels Bohr

THE STRUGGLE between God and the devil, the Lord and Mephistopheles, is the core of the Faustian drama. There was no doubt in the minds of the young physicists gathered in Copenhagen that the Old Man, the Lord, was Bohr. It was also clear that Pauli's sharp mixture of humor and derisiveness meant he should be Mephisto (as he was called in the skit), the devil, *Faust*'s great ribald figure. The "Copenhagen Faust" makes sure the audience knew at once the devil's identity by having lines from the prologue changed to

From time to time it's pleasant to see the dear Old Man,
I like to treat him nicely—as nicely as I can.
He's charming and he's lordly, a shame to treat Him foully—
And fancy!—he's so human he even speaks to Pauli!

It was clear that the long-standing relationship between Bohr and Pauli was filled with admiration, affection, and respect. Many physicists who knew Bohr testified that the letters he received from Pauli were the most treasured. Leon Rosenfeld, one of the young physicists (incidentally he played the role of Pauli/Mephisto in the 1932 performance), recalls what it was like when one came in the mail:

> The arrival of a letter from Pauli was quite an event: Bohr would take it with him when going about his business, and lose no occasion of looking it up again or showing to those who would be interested in the problem at issue. On the pretext of drafting a reply, he would for days on end pursue with the absent friend an imaginary dialogue almost as vivid as if he had been sitting there, listening with his sardonic smile.

Pauli/Mephisto speaking to Bohr/the Lord.

Bohr and Pauli's friendship began in 1922 during the course of the seven lectures Bohr gave in Göttingen that year, the same set during which Bohr first met Heisenberg. At the time Bohr was looking for a young collaborator to help him prepare a German edition of his works. He asked Pauli if this would interest him or if it might pose difficulties. Pauli, nondeferential as usual, replied that Bohr's physics would be absolutely no problem for him, child's play, but learning Danish was an entirely different matter, probably beyond his mental capabilities. Bohr burst out laughing, and Pauli joined in. The friendship was sealed. Three months later Pauli went to Copenhagen for the first time. Incidentally, he also learned Danish.

Niels Bohr, this fortunate man, was born in Copenhagen in 1885 to

Christian Bohr, of old and prominent Danish ancestry, and to Ellen Adler, the daughter of a Jewish banking family that had immigrated to Denmark from Germany. The couple, warm, loving, and cultured, had a daughter, Jenny, and then two boys, Niels and Harald. The two brothers, only a year and a half apart, were inseparable, a closeness that remained throughout their lives. Both boys were very good athletes—Harald was even a member of the 1908 Danish Olympic soccer team—but Niels was more capable at manual tasks. Harald, later a world-famous mathematician, would laugh about one way the two differed markedly. Known as a brilliant lecturer while Niels was a notoriously poor one, Harald explained the difference: "At each place in my lecture I speak only about those things which I have explained before, but Niels usually talks about matters which he means to explain later."

The two also showed a difference in how they approached mathematics and science. Harald was a solitary thinker, while Niels, from a young age on, felt the need for a sounding board to clarify his notions, testing and retesting them on a companion. He would pause in this, swaying to and fro, lost in concentration, his face going blank with jaw drooping as he focused intently on the problem at hand. As the answer came to him, his face would break into a big smile, and the discussion could then continue. His mother told the story of once overhearing somebody taking pity on her for having such a dumb-looking child, but it was only Niels absorbed in thought.

Bohr was a strong, big-boned man with a large head, oversized hands, and protruding teeth—not conventionally handsome, but ruggedly attractive, looking more like a fishing-boat captain than a professor. He possessed enormous physical and mental energy and in his fifties was still exhausting physicists half his age. He would walk, talk, and smoke his pipe, pausing frequently to relight the bowl. It didn't matter whether his interlocutor was Einstein or a student: Bohr pursued the conversation with the same steadiness, gentleness, and unwavering quest for answers. Contrasting with the large physical presence, his voice was soft and gentle.

When working on a paper, which was most of the time, Bohr would select an assistant from among the young physicists in Copenhagen. This assistant, affectionately dubbed the victim, was supposed to sit in place while Bohr paced around the room, constantly puffing away at his pipe, working and reworking his ideas, talking aloud as the idea took shape, trying and retrying to dictate his sentences to the victim. In the middle of the day the victim would usually have lunch with Niels and Margrethe, and then work would continue. Sometimes it was hard. Victor Weisskopf, Bohr's victim in the fall of 1932, recalled having problems controlling his own lifelong habit of pacing while working: "Only one of us is allowed to move was his rule so I sat there day after day in agony." But he added that, despite the suffering, it had been a great privilege to spend so much time with Bohr and to observe at close quarters his thought processes.

Of course the charmed Copenhagen atmosphere would have been impossible without the regal and beautiful Margrethe. She was by Niels's side from 1916 on, always smoothing the way for him, taking care of their six sons, hosting with a smile the stream of visitors both in Copenhagen and at their simple seaside house in Tisvilde. She also freed Niels to go for his day or weeklong trips with young physicists or to simply spend time undisturbed in his study, together with his victim.

Whether there, on a walk, or in a lecture hall, Bohr was invariably polite, interrupting others only with his usual phrase, "I don't mean to criticize, only to learn." He was, however, relentless in trying to clarify physics issues that troubled him, never more so than during 1926 and 1927, the years in which the interpretation of quantum mechanics was hammered out.

The framework for understanding this magnificent theory was in great part built in Copenhagen with Bohr and Heisenberg as the chief architects, but the intensity of discussions on the subject was all-consuming in the world's major physics centers as scientists fiercely debated the pros and cons of the two competing formulations that suddenly appeared, Heisenberg's matrix mechanics and Schrödinger's wave mechanics. Seeming at first to be altogether unalike, the two were soon shown to be essentially identical in their predictions of experimental results, but Schrödinger's version was couched in a mathematical language the scientific community found more familiar and therefore more accessible. However, each theory's creator had a completely different interpretation of the mathematical formalism: Heisenberg believed the electron paths around the nucleus, the so-called orbits, were unobservable and therefore meaningless, while Schrödinger maintained that these paths, far from meaningless, were determined by the behavior of the waves he proposed as the basis of his mechanics.

Bohr realized that understanding what either formulation meant was still an open question. Seeing this dilemma as the key to making further progress, he invited both Heisenberg and Schrödinger to come to his institute. Schrödinger arrived in Copenhagen in October 1926, having already swayed Berlin, Munich, and Zurich audiences with his own point of view. This time the reception was less sympathetic. Heisenberg, who was present at many of the discussions, provides the following recollection in his memoir *Physics and Beyond:*

> The discussion between Bohr and Schrödinger began at the railway station in Copenhagen and was carried on every day from morning until late at night. Schrödinger stayed in Bohr's house and so for this reason alone there could hardly be an interruption in the conversations. And although Bohr was otherwise most considerate and amiable in his dealings with people, he now appeared to me almost as an unrelenting fanatic, who was not prepared to make a single concession to his discussion partner or to tolerate the slightest obscurity. It will hardly be possible to convey the intensity of passion with which the discussions were conducted on both sides, or the deep-rooted convictions which one could perceive equally with Bohr and with Schrödinger.

Bohr insisted that electrons were, in a sense, jumping from one orbit, or as he called it, one quantum state, to another. Heisenberg remembers the exasperated Schrödinger saying to Bohr, "If all this damned quantum jumping were really here to stay then I should be sorry I ever got involved with quantum theory," to which Bohr then replied, "But the rest of us are extremely grateful that you did." Nothing stopped the discussions. Finally Schrödinger, exhausted and feverish, took to his room. Margrethe Bohr administered to him, bringing him tea and cake, but Niels Bohr would not let go. Sitting on the edge of Schrödinger's bed, he continued to argue with him. As they grappled together, neither managed to persuade the other of his views, but they came away from their discussions with a newfound admiration and respect for each other. One can only imagine how many times during these conversations Bohr uttered "I don't mean to criticize, but . . . " followed by "I don't quite agree" or "Please try to see it this way."

That phrase of Bohr's, "I don't mean to criticize," became so standard that young physicists in Copenhagen started saying to one another, "I don't mean to criticize, BUT . . ." In the Copenhagen parody of *Faust*, Bohr/the Lord exits with the lines

> *I say this not to criticize, but rather just to learn. . . .*
> *But now I have to leave you. Farewell! I shall return!*

And return he did, again and again, to points that bothered him, just as he had with Schrödinger. Incidentally, those discussions were usually settled in Bohr's favor.

Paul Ehrenfest

WHILE BOHR was portrayed as the Lord and Pauli as Mephisto in the skit, the part of the troubled, conflicted Faust, torn between the two of them, was portrayed as a takeoff of Paul Ehrenfest. Delbrück, the casting director as well as the author of the play, certainly did not appreciate the depth of Ehrenfest's anguish at the time, but perhaps he sensed enough of it to make the identification. Nobody knew until after his death that

Ehrenfest/Faust in his study.

Ehrenfest had composed a letter a few months after the 1932 meeting in which, hinting at suicide, he described his feelings of inadequacy. He had at first intended to mail it to a few friends, including Bohr and Einstein, but afterward changed his mind, never sending it to anybody.

Paul Ehrenfest, short, with brush-cut hair, a mustache, and bright eyes framed by round eyeglasses, was born in Vienna in 1880, making him one year younger than Einstein and five older than Bohr. His parents were among the large number of Jews who had flooded into the capital of the Austro-Hungarian Empire after the Emperor Franz Josef lifted the restrictions that had confined Jews to ghettos, barred them from many professions, and limited their children's educational opportunities. On entering the mainstream, many of these families also renounced their Judaism, officially or unofficially, regarding it as a relic of their provincial past and possibly an obstacle to career advancement. Some converted to Protestantism, some to Catholicism, and others simply chose to regard themselves as agnostics.

Ehrenfest's parents lived in the predominantly working-class Catholic district of Favoriten, where they owned and managed a grocery store. The shop was downstairs, and the family lived above it, in a ménage that included Paul, four older brothers, two maids, a nursemaid, and six shop assistants.

His father's favorite, Paul was an afterthought, eight years younger than the youngest of his four brothers. Being nervous and sickly, lacking the group protection the four older brothers provided for one another, he suffered a difficult childhood, made worse in his Catholic neighborhood by being continually taunted as a Jew. The engendered insecurity was heightened by the death of Ehrenfest's mother when he was ten and of his father when he was sixteen. Left largely in the care of his brothers, he discovered a gift for mathematics and science that carried him along, though the dark moments of his youth reappeared throughout his life. Visiting Lake Como in Italy as a young man, he described in his diary a joy he felt at seeing the great sights, but then added, "Isn't the feeling of nervous disgust that is so familiar to me now just another form of the same psychological disposition that used to appear so often to me as Sunday boredom? How to cure it?" There never would be an answer.

Ehrenfest entered the university in 1899. However, German and Austrian university students often spent only a year or two at one university and then went off to another one, their so-called *Wanderjahre* (wander years). Following that tradition, Ehrenfest moved from Vienna to Göttingen in 1901. There he met and fell in love with a young Russian woman physicist a few years older than he. By the winter of 1902–03, they decided to get married and relocate in her native country.

This was not so simple, because the laws of the Austro-Hungarian Empire at the time forbade marriage between a Jew and a Christian. By officially forswearing all religion, they cleared that hurdle; in 1907 the young couple finally moved to St. Petersburg, the capital of Tsarist Russia. Before then, they returned briefly to Vienna for Ehrenfest to obtain his doctoral thesis. Doctorate or no doctorate, very few academic positions were available in Russia and certainly none for a foreigner who had forsworn religion, so the Ehrenfests eked out their existence through a bit of family money and some temporary teaching jobs.

By the winter of 1912, Ehrenfest was miserable, isolated intellectually, with no job in sight. He decided to take a two-month tour of German-speaking physics centers to see if he could find a position. Leipzig, Berlin, Munich, Vienna, and Zurich were on his route, and he stayed whenever possible with friends in order to save money. On his homeward trip he stopped in Prague to visit a young physicist he had corresponded with briefly but never met: Albert Einstein.

From the moment Einstein greeted him at the train station, the two of them became embroiled in conversations about physics. What had been intended as a short visit stretched out to a week. When they weren't struggling with some problem in relativity, statistical mechanics, or quantum theory, the two played Brahms sonatas for violin and piano or went for walks through the streets of Prague. They were about the same age, both Jews, both music lovers, both from nonacademic backgrounds, both unconventional. As Einstein later remembered, "Within a few hours we were true friends—as though our dreams and aspirations were meant for each other."

Einstein, planning to leave Prague at the end of the year to take up a professorship in Zurich, suggested that his new friend fill his soon to be vacated position. Once again there was a hitch: professorships in the Austro-Hungarian Empire were only open to individuals with a religious affiliation. Einstein, having cleared this hurdle by listing himself as a follower of the "Mosaic creed," suggested Ehrenfest do the same, but he refused. He would not go back on his earlier renunciation of religion. In April Einstein wrote him, "I am frankly annoyed that you have this caprice of being without religious affiliation; give it up for your children's sake. Besides, once you are a professor here, you can go back to this curious whim again—and it is only necessary for a little while." Despite Einstein's entreaties, Ehrenfest would not change his mind.

Even though Ehrenfest hadn't secured a position, the trip was a success. He made many new friends, and word spread that he was a young man with extraordinary talent. That summer Ehrenfest received a letter from Hendrik Lorentz, a universally admired scientist and a winner of the 1902 Nobel Prize in Physics, the second ever awarded. (Wilhelm Röntgen had received the first for his discovery of X-rays.) Fluent in the major European languages and a skilled diplomat, Lorentz mediated physics arguments and presided over the prestigious Solvay conferences, subtly directing discussions and moving agendas forward. He had filled for over thirty years the chair of theoretical physics at Leiden, the oldest university in the Netherlands, known since the seventeenth century as a center of tolerance as well as one of Europe's prominent seats of learning. Now, although only in his late fifties, Lorentz wanted to give up the professorship to devote himself to research and other activities.

Before retiring, Lorentz felt he had to be sure his beloved Leiden physics institute would be in good hands. He wrote to Einstein, asking if he would consider taking the post, but Einstein had already committed himself to moving from Prague to Zurich. Einstein later told a friend that he was glad the decision had been made because he preferred Zurich to Leiden but would not have been able to say no to Lorentz.

In the meantime Lorentz had read some work by Ehrenfest and his wife, herself a distinguished theoretical physicist. The topics treated were familiar to him, and he was very impressed by the depth of the discussion and the originality of the views. On the lookout for new talent and with Einstein out of the picture, Lorentz tried to find out more about the young couple. Never having met either, he wrote Arnold Sommerfeld for an evaluation of Ehrenfest's persona, trusting his judgment and knowing that Sommerfeld had recently interacted with him. Lorentz may well have heard from Einstein as well, but Sommerfeld was Lorentz's contemporary and probably better able to judge a younger physicist's ability as a teacher.

The answer from Munich came back quickly. Ehrenfest was portrayed as a very promising scientist, a man of principle, and a great teacher. "He lectures like a master. I have hardly ever heard a man speak with such fascination and brilliance. Significant phrases, witty points and dialectic are all at his disposal in an extraordinary manner." This was enough for Lorentz. He sent Ehrenfest a letter, expressing an interest in bringing him to Leiden and asking him to please wait before accepting any other offers.

On reading the letter, Ehrenfest was overwhelmed with emotion. He had thought there were no open positions for him, and now, beyond his wildest dreams, he was being considered as Lorentz's successor. Admittedly it would require learning Dutch, but that was a small price to pay for such an unbelievable honor.

The offer of the Leiden chair materialized quickly. Accepting it immediately, the Ehrenfests arrived in Leiden in October 1912. Teetotalers, smoke free, and vegetarian, they were eccentric by the standards of the day. They also decided that their three children should be schooled at home rather than in the public schools. The life that came with these convictions was noisy and chaotic, with nursemaids, tutors, frequent visitors, students, and scientists from abroad roaming through the house. Einstein, who in 1914 moved from Zurich to Berlin, came frequently for a few days or a week to discuss physics with them and with Lorentz; he enjoyed the unconventional atmosphere. It was a stark contrast from his own marriage, which had essentially ended in 1914, and from his difficult relationship with his own children.

Einstein and Ehrenfest would walk, talk, and write on the study's blackboard; from time to time they moved to the music room to play. By special dispensation Einstein was allowed to smoke in his third-floor bedroom. In 1919, writing from Berlin after one stay at their house, Einstein observed, "I have never before taken part in such a happy family life. It proceeds smoothly from two independent people, who are not united just by compromises. I have come to feel that all of you are a part of me and that I belong to you."

As Ehrenfest threw himself into teaching, his reputation grew. Physicists in Leiden and abroad looked to him for his ability to glean a problem's essence, listening for his oft-repeated phrase *Was ist der Witz?*, roughly translated as What does this really mean? or What are you getting at? This inquiring spirit often crossed over into personal matters, and at times he delved into the details of friends' and students' lives. Most loved it, but one described him as sucking the lifeblood out of them.

Problems like this had already occurred earlier. Back in Vienna, when they were university students together, Ehrenfest and Meitner had never become close friends despite their common interests. Martin Klein, Ehrenfest's distinguished biographer, wrote to her, asking why this was so. He summarized her answer: "She found Ehrenfest's need to probe into the very soul of anyone with whom he came into contact was too much for her, and it prevented a very close acquaintance."

Ehrenfest was not too much for Einstein, and he was also not too much for Bohr. Bohr and he began taking note of each other's work during the 1914–16 period, but World War I kept them apart. However, Hendrik Kramers, Bohr's Dutch assistant, managed to travel back and forth between Denmark and the Netherlands. When the war was over and it was time for Kramers to defend his thesis in Leiden, Bohr came to attend the ceremony and give a lecture. Finally meeting, Ehrenfest and Bohr quickly started a mutually treasured friendship. When Bohr left, Ehrenfest wrote him, "You had gone, the music had faded away." Two years later Ehrenfest went to Copenhagen to lecture.

Despite Ehrenfest's growing reputation, the doubts and dark moments that plagued him since childhood never went away. His diaries and even his letters to good friends often reflect this deep sense of inadequacy. Writing Einstein in 1920, he said, "What I can do is not science but only a bit of entertaining salon conversation or promenade conversation—the physics done by others." That was certainly not the way those others perceived him or even the way he thought about himself in his better moments. Trying to console him, Einstein wrote back, "Don't complain and don't vex yourself. We too may make use of the human law that one gets stupider as one grows older. In that way we acquire the merit of easing the conscience of others." Concerned that he might have irritated Einstein, Ehrenfest replied, "Don't be impatient with me. Bear in mind that I hop around among all of you big beasts like a harmless and helpless frog who is afraid of being squashed."

At forty, Ehrenfest had been feeling old. As the 1920s came to an end and he turned fifty, his mood seems to have become more consistently dark. Part of it may have been age, but the vertiginous pace of physics after the 1925–27 quantum mechanics revolution was putting a great mental strain on all the participants. Unlike Einstein, Ehrenfest accepted the Copenhagen interpretation of quantum mechanics, but unlike Bohr, he wasn't able to lead it into new territory.

Lise Meitner

THE ONLY ONE of the seven physicists from the front row not parodied in the "Copenhagen Faust" was Lise Meitner. There was a good reason for this. She was an experimentalist, and this was a skit written by theorists, intended primarily as a vehicle for making fun of other theorists. This did not mean that experiment was ignored. Quite the contrary was true, so much so that Pauli/Mephisto, near the end of the skit, offers this reminder to the audience:

That which experiment has found—
Though theory had no part in—
Is always reckoned more than sound
To put your mind and heart in.

One might even go a step further, asserting that at times Meitner the experimentalist would have to be the arbiter, settling which of two competing theories was correct, perhaps even adjudicating victory to either Bohr/the Lord or Pauli/Mephisto.

Lise Meitner's career is a parable of one scientist overcoming seemingly insurmountable obstacles. Like Pauli's father and Ehrenfest's parents, her father and mother were also part of the great migration of Jews from the provinces to Vienna during the second half of the nineteenth century, a migration that enriched the city's cultural life with the likes of the author Arthur Schnitzler, the composer Arnold Schoenberg, and of course Sigmund Freud. Lise's own father, Philipp, became one of the city's first Jewish lawyers.

Unlike Ehrenfest's parents, who moved into a Catholic neighborhood, the Meitner family settled in Leopoldstadt, a section of Vienna that had once been a ghetto and had remained largely Jewish in character. It was now growing by leaps and bounds as immigrants moved into the houses along the tree-lined avenues that ran from the city center to the beautiful Prater, the park on the district's northeastern edge. The Leopoldstadt apartments were usually spacious, the households had servants, and the salons featured the inevitable grand piano that played such a large role in the life of musically conscious Vienna.

Families were typically large; the Meitners had eight children, three boys and five girls. All eight pursued higher education, surprising when one realizes that even the special schools that prepared students for entrance to university were closed to women until 1897. Lise was born on the seventh of November, 1878, a few months before Ehrenfest. Forced to delay her preparations until the ruling against women had been reversed, Lise didn't enter the University of Vienna until 1901. She received her doctorate in 1906, staying then for an extra year to pursue research in the newly emerging field of radioactivity.

Given the limitations on women's careers at the time, Meitner felt the best position she could obtain, even with a degree in hand, would be one teaching science at a girls' school, and that would mean giving up research. Strongly interested in pursuing her research career at least for a little longer, she asked her parents to subsidize a one-year stay in Berlin. They agreed. On arrival in the German capital in the fall of 1907, Meitner called on Heinrich Rubens, the senior professor of experimental physics, the same man who had started quantum theory by asking Planck to find a formula that would explain his data. Meitner inquired if he could find a laboratory for her in which she could pursue the investigations she had begun in Vienna. He introduced her to a young man who was looking for somebody to work with him.

Otto Hahn, the same age as Meitner, had not had a distinguished university record. Leaving Germany for England in 1904 to prepare for a commercial career in chemistry, Hahn was instead swept up into radioactivity research. Having decided to change course because of his interest in this new field, he tried to pursue his inquiries for as long as it was possible. Some initial successes got him an appointment with Rutherford at McGill University and later a position as an *Assistent* in the laboratory of Emil Fischer, Berlin's great organic chemist. However, *Assistent* was the bottom of the German academic ladder at the time, and furthermore nobody in Berlin was working in the area of radioactivity. Fischer, who liked to say that the human nose was the most sensitive chemical instrument known, was neither interested nor encouraging about Hahn's chosen line of research. Nor of course is the nose the appropriate instrument for studying radioactivity.

Seeing it as a possible entry into the mysteries of what lies inside atoms as well as a new source of energy, physicists seemed to find radioactivity more interesting than did chemists. Searching for intellectual companionship, Hahn began attending their weekly colloquia, slowly getting to meet like-minded scientists. Hearing Hahn's wish to find a co-worker and knowing that Meitner and he had similar interests, Rubens arranged a meeting. They liked each other and agreed on how to proceed; a match was made.

Beginning in Berlin was even harder for Meitner than for Hahn, because Fischer's Chemical Institute was off-limits to women. After negotiations, a compromise was reached whereby Meitner could use a basement carpenter's room that had a separate entrance to the street, but she was not allowed to go upstairs into the institute, even to Hahn's laboratory. If she had to use a bathroom, she needed to walk to a nearby restaurant. When the Rutherfords spent a few days in Berlin on their way back to Manchester after the 1908 Nobel Prize ceremonies, the men talked in Hahn's laboratory while Meitner went shopping with Mary Rutherford. Meitner was also unpaid, which meant continuing to live in a furnished room on an allowance from her parents. But with the exciting work propelling her forward, she extended her stay in Berlin well beyond what either she or her parents had originally envisioned.

Hahn and Meitner were an odd combination, not least because of the rarity of women in research science, but their union was both happy and successful. Starting in the fall of 1907, neither of them yet thirty, they continued working together for over thirty years, only stopping their collaboration when Meitner, in extremis, fled Germany, endangered because of her Jewish origins. Their partnership had what seems today like a curious formality. They never had a meal together, except on official occasions, never went for a walk together, and never accompanied the other home at night. They did not even use the informal pronoun *du* (you) with each other until well into the 1920s. Nevertheless each considered the other a good friend, jolly and supportive.

Despite the pair's progress in research, the chemists in Fischer's institute continued to be unwelcoming, but fortunately for Meitner the physicists took note of her and began trekking over to the carpenter's shop to talk. One of them in particular, a very prominent figure, took a strong interest in her career. It was Max Planck.

Meitner admired Planck tremendously. Fifty years after their first meeting, she wrote,

> He had an unusually pure disposition and inner rectitude, which corresponded to his outer simplicity and lack of pretension . . . Again and again I saw with admiration that he never did or avoided doing something that might have been useful or damaging to himself. When he perceived something to be right, he carried it out, without regard for his own person.

Meitner had been attending Planck's lectures and soon had become a friend of the whole Planck family. She went for walks in the country with his twin daughters and to musical evenings at their house. Concerned about her, particularly after her father's death in 1910, Planck did something extraordinary, a move Meitner always regarded as the turning point in her career. In 1912 he appointed her as his *Assistent*, her first paid position and the first woman *Assistent* in Prussia. Within a year Fischer followed Planck's lead, making sure she received a rank comparable to Hahn's. By 1913, thanks to Planck's and then to Fischer's interventions, Hahn and Meitner were both salaried, had research funds and a four-room laboratory in the north wing of the new Institute for Chemistry of the prestigious Kaiser Wilhelm Society. Founded only a few years earlier, the society's aim was to advance German research and development in science.

Planck's action seems characteristic of the man. In his biography of Einstein, Abraham Pais has a preface that concludes, "Were I asked for a one-sentence biography of Einstein, I would say, he was the freest man I have ever known." Nobody would ever dream of applying this phrase to Planck, conventional in many ways, deeply tied to Germany and its institutions. At the same time, he continually questioned these loyalties and often struggled with their implications, certainly during World War I and again as Nazism arose in his country. Meitner was correct in emphasizing how Planck always tried to do what he thought was right, without regard for his own welfare. It is not by chance that John Heilbron's biography of Planck is entitled *The Dilemmas of an Upright Man.*

One cannot imagine Einstein writing about women, as Planck did in 1897, that

> Amazons are abnormal, even in intellectual fields. In certain practical situations, for example, women's health care, conditions might be different, but in general it cannot be emphasized strongly enough that Nature itself has designated for woman her vocation as mother and housewife, and that under no circumstances can natural laws be ignored without grave damage, which in this case would appear especially in the next generation.

On the other hand, one also cannot imagine Planck behaving toward women as selfishly as Einstein sometimes did, nor would Einstein have worried about Meitner's future. Again, as Meitner said of Planck, "When he perceived something to be right, he carried it out, without regard for his own person."

Hahn also became a friend of the Planck family. With a strong tenor voice, he was welcomed at musical soirees, though he never felt as much at ease as Meitner did. They had attained similar academic status, but Meitner's father was a lawyer while Hahn was the son of a peasant who had worked his way up to being a storeowner; Hahn never quite forgot that. As Hahn recalls in his memoirs, "All these people belonged to good families, and it was not easy to get accepted in those circles."

Those musical evenings were infrequent examples of the two interrupting their intense work schedule. Early Monday morning Hahn, the chemist, and Meitner, the physicist, were always back in the lab. Though their early training and inclinations were different, Meitner was also an excellent chemist and Hahn an equally adept physicist, the duo's distinct but complementary ways of approaching problems being one of their great strengths.

Despite her social graces, Meitner never married, nor did she ever have any known relationships that might have led to marriage. A daughter of a friend once asked her about this, given that she was beautiful and surrounded by young men, her fellow scientists. Lise laughingly replied, "I never had time for it." Perhaps she might have married had she developed early on the kind of personal and professional partnership that Marie Curie had with Pierre, but the only candidate was Otto Hahn, and the two of them were an unlikely pair for many reasons. Later in life, once she had established her position, it would have been hard to imagine her in a typical German marriage.

By 1932, now in her early fifties, such thoughts were a thing of the past if she had ever entertained them. With her career well established, Meitner had moved from her furnished room to a graceful apartment near her laboratory. She was now surrounded by friends and had a rich social life, but as before, science remained the central focus of her life.

One particularly dear and close friendship of Meitner's was the one she had with Bohr. They met in 1920. The construction of Bohr's institute had been occupying him for some time, but when he received an invitation that year to speak in Berlin, he knew there were many reasons he should accept, foremost among them that he had never met Planck or Einstein. Before the start of the war, Bohr was a relative unknown—now, at age thirty-five, he was among the greats. Planck, Einstein, and Bohr, all three extraordinary, were increasingly being regarded as the pioneers of the emerging quantum theory. Physics Nobel Prizes arrived for them in order of age: Planck in 1918, Einstein in 1921, and Bohr in 1922.

Planck, considerably older than Bohr, was no longer doing research at the cutting edge, but for the Dane, simply getting to know him would be a privilege. But Einstein, only six years older than Bohr, was very much involved in the mysteries of quantum physics, so Bohr planned to have significant exchanges with him about the direction the field might be heading. Though the two of them had not previously corresponded, they had followed each other's work and were both good personal friends of Ehrenfest's.

Surpassing all his expectations, Bohr's reception from Einstein was extraordinary. On returning to Copenhagen, he found a letter from his new friend:

> Not often in life has a person, by his mere presence, given me such joy as you. I now understand why Ehrenfest is so fond of you. I am now studying your great papers, and in doing so—when I get stuck somewhere—I have the pleasure of seeing your youthful face before me, smiling and explaining.

Deeply touched, Bohr replied:

> It was one of the greatest experiences ever to meet you and talk with you and I cannot express how grateful I am for all the friendliness with which you met me on my visit to Berlin. You cannot know how great a stimulus it was for me to have the long hoped for opportunity to hear of your views on the questions that have occupied me. I shall never forget our talks.

Einstein was not the only new friend Bohr acquired during that April 1920 visit to Berlin. Others were eager to meet him. Einstein was a great man, but he kept largely to himself, whereas Bohr seemed so approachable, so friendly, and so happy to talk with junior researchers. Lise Meitner, still establishing herself, knew that at the weekly Berlin physics colloquium the famous professors seated in the front row—in Germany they were nicknamed Bonzes, bigwigs in modern usage—dominated the question-and-answer period. Half-jokingly she organized what came to be known as the Bonzenfreie (Bonze-free) Kolloquium, a day with Bohr where Bonzes would not be invited. Bohr accepted with pleasure, and we have as a souvenir of that occasion a picture of thirteen formally attired youngish men surrounding Bohr and a pleased Lise Meitner (see photograph 5). A year after her first meeting with Bohr, Meitner went to lecture in Copenhagen, beginning a long and happy relationship with the whole Bohr family.

Chapter 5

The Front Row: The Revolutionaries

Age is, in sooth, a fever cold,
With frost of whims and peevish need:
When more than thirty years are told,
As good as dead one is indeed:
You it were best, methinks, betimes to slay.

Goethe, *Faust, Part II,* act 2, 220–24

By Copenhagen meeting standards, Bohr, Ehrenfest, and Meitner were all members of the older generation, old enough to have puzzled about the atom even before the discovery of the nucleus. The three had seen the revolution of quantum mechanics come and had played their parts. What more could they now achieve? It was hard to escape the realization that they were some twenty years older than most of the others at the gathering. Perhaps Meitner, just over fifty, didn't feel the difference quite as much because she was an experimentalist and that was a different story. Those who pursued that career seemed to blossom later and last longer, which made sense because it took them years to acquire their laboratories and tools. The familiarity gained through long and patient work with a given kind of apparatus paid off in ways that did not have a counterpart in theory. Even though it was also important for experimentalists to learn new techniques to stay at the cutting edge of research, experience seemed to count for more in their field than it did in theoretical work.

The whole Copenhagen group knew that the young led the quantum mechanics revolution, but by 1932 even the three most esteemed members of the new generation were no longer youthful. Pauli at thirty-one, Heisenberg at thirty, and Dirac at twenty-nine might well have remembered what the student, Baccalaureus, says in *Faust*, the lines quoted above as the epigraph to this chapter.

Old Age Is a Cold Fever

SITTING IN THE FRONT ROW at the 1932 Copenhagen meeting, Heisenberg reflected on having turned thirty a few months earlier. It was reassuring in some ways to have next to him Bohr and Ehrenfest, the latter just over fifty years old and still active. However, the professor from Leiden, though as sharp as ever, had not been significantly productive in many years, and even Bohr was not quite the innovator he had once been. Heisenberg might have wondered why this was so. Of course, as you age, responsibilities increase and energy levels decrease, but there was something else to it. Maybe you stop believing you can change the world, or maybe you realize more quickly that a crazy idea is just that and do not pursue it as vigorously. Maybe you can't assimilate new material as rapidly as you did when you were younger. Perhaps, having set a new course for physics once, you experience a psychological and intellectual resistance to shifting direction again. But Heisenberg, still extraordinarily quick and mentally strong, was confident that his career in physics was far from over.

A new generation of theorists was now coming along. Turning around in his seat, Heisenberg could see his own student Carl Friedrich von Weizsäcker, only twenty years old but so brilliant and already so knowledgeable. Then there was Felix Bloch, who had been his first graduate student and an extraordinary one at that. Now in residence at the Copenhagen institute, he was sitting in the row behind Carl Friedrich. It was a source of pride to have such wonderful students, but it also meant he, Heisenberg, was not the prodigy anymore.

As quantum mechanics was being developed in the 1925–27 period, the Göttingen group had jokingly given it the name *Knabenphysik* (boy's physics) because the leaders all seemed to be *Knaben,* one younger than the other—Pauli, Heisenberg, Dirac. But now, in 1932, those *Knaben* were either approaching or just past thirty. Their achievements had been magnificent, but might their best days in research be behind them?

In case they had forgotten Baccalaureus's lines, the finale of the "Copenhagen Faust" had the onstage counterparts of Heisenberg, Dirac, and Pauli reinterpreting them. Heisenberg speaks first:

> *Insofar as we have won half the world,*
> *What have you done? Nodded, sunned yourself*
> *Dreamt, woken up, plans and always plans!*

Echoing Baccalaureus, Dirac adds,

> *Certainly! Old age is a cold fever*
> *That every physicist suffers with!*
> *When one is past thirty,*
> *He is as good as dead!*

Making the message even clearer, Heisenberg suggests what should be done with the over-thirties:

> *It would be best to give them an early death*

Finally Pauli, who as Mephisto has been the most loquacious of the three throughout the skit, concludes it. At a loss for words and depressed by the realization of his own aging, he brings down the curtain with

> *Pauli has here nothing more to say!*

Sadly, perhaps predictably, though each one of the three continued to be productive, nothing they achieved after 1932 had the same impact as their earlier work. It had been a magnificent run.

Finis—The skit has come to an end.

Werner Heisenberg

SUPERFICIALLY, the circumstances of Heisenberg's life sound very much like Bohr's: a devoted mother, a father who was a university professor (physiology in Bohr's case, Greek in Heisenberg's), a single brother close in age, and a lifelong interest in outdoor activities and athletics. Professional success, recognition, and acclaim came early to both, even sooner to Heisenberg than to Bohr. Their personal lives were fortunate as well: fruitful marriages and many children (six for the Bohrs, seven for the Heisenbergs). Finally death arrived at the same point in their life trajectory (Bohr was seventy-seven and Heisenberg seventy-five). But these parallels distract from some dramatic differences between them.

They each had a strong attachment to their native countries, but Bohr's devotion to Denmark, his efforts throughout his life to help his small native land, and his heroic actions to protect Danish Jews during World War II have received universal praise, whereas Heisenberg's acquiescence to the Nazi regime and his command of German nuclear research in wartime have tarnished his human image, irremediably in the eyes of some. The 1941 meeting in Copenhagen between Bohr and Heisenberg, a flash point for the controversy about their wartime activities, highlighted the contrast in their stances toward their respective roles during that war.

On a personal level Bohr's life appears blessed from childhood onward in a way that Heisenberg's never was. Bohr's brother, Harald, remained his closest, his most intimate friend throughout his life. When in 1912 Niels first began to uncover the atom's secrets, he immediately wrote his brother:

> Dear Harald:
>
> Perhaps I have found out a little bit about the structure of atoms. Don't talk about it to anybody, for otherwise I couldn't write to you about it so soon. If I should be right . . .
>
> This was intended only as a little greeting from your Niels who is very much longing to talk with you.

There was no comparable contact between Erwin and Werner Heisenberg. Their father, who had struggled to obtain his university position and was keen on preserving the family's upward mobility, encouraged competition rather than cooperation between the brothers. They responded ferociously, but in the process became increasingly estranged. As adults they were cool with one another, exchanges between them taking place only on rare family occasions.

Nor does Heisenberg's later married life seem to have the joyous quality that appears so evident in descriptions of Niels and Margrethe. At age thirty-five, Werner, lamenting his state to his mother, wrote her, "The single life is bearable to me only through my work in science, but for the long term it would be very bad if I had to make do without a very young person next to me." A few months later he met and quickly married twenty-two-year-old Elisabeth Schumacher. The marriage was seemingly a happy one, but perhaps without as much intimacy as both may have wanted. As Heisenberg's son Jochen said to Michael Frayn in an exchange about the latter's play *Copenhagen,* "I never saw my father express emotion about anything, except music. But I understand that the characters in a play have to be rather more forthcoming than that."

Perhaps Heisenberg felt more strongly than Bohr the need to be guarded with his feelings. The chaos of World War I certainly had affected him. In high school at the end of the war, he described his thoughts at the time in his 1969 book of reminiscences:

> It must have been the spring of 1920. The end of the First World War had thrown Germany's youth into great turmoil. The reins of power had fallen from the hands of a deeply disillusioned older generation, and the younger ones drew together in larger and smaller groups to blaze new paths, or at least to discover a new star to steer by.

Heisenberg found two stars to follow. One was a youth movement he belonged to for many years, the Neupfadfinder (New Pathfinders), an all-male brotherhood that emphasized notions of trust, purity, morality, and camaraderie, all steeped in a romanticism that harked back to early Teutonic tales. The idea was not altogether original, the New Pathfinders being direct heirs of groups such as the Wandervogel (Migratory Bird) that had flourished in pre–World War I Germany. Heisenberg and his comrades took long camping, hiking, and mountaineering trips; they sang, recited poems to one another, and talked of a new German Reich (Empire) that would embody these ideals. Imbued with a sense of pan-German nationalism, the Pathfinders traveled to neighboring countries in order to express their solidarity with former Germans now under foreign rule. The South Tyrolean Alps and the Dolomites, ceded to Italy as part of the peace settlement that followed World War I, were for them a particularly painful example of annexation by foreign cultures. Once Austrian, these mountains remained a favorite stomping ground for German hikers, mountaineers, and vacationers, especially since the lingua franca of most of the region's villages remained their ancestral German.

In retrospect many of the Neupfadfinder's activities sound at best naïve and at worst proto-Nazi, but one has to remember the times and the milieu in which those youths grew up. Germany had suffered a disastrous defeat in World War I, the country was in shambles, the political and economic situations were chaotic, and dishonesty was rampant. The pathfinders yearned nostalgically for the ideal of an older, imagined, uncorrupted Germany. It is also worth mentioning in their defense that Heisenberg's immediate group seems to have veered more toward the church than to the Nazi Party and did not show signs of the growing anti-Semitism that was spreading through Germany.

Heisenberg's other star was of course science. And though both stars were important in his formative years, he kept them separate, alternating periods of intense mental work with long sessions of relaxation with his old comrades. Years later, well after his Neupfadfinder days were over, he still kept this pattern of following sessions of intense work with long stays in the mountains. Describing his student days, Heisenberg wrote,

> My first two years at Munich University were spent in two quite different worlds: among my friends of the youth movement and in the abstract realm of theoretical physics . . . Both worlds were so filled with intense activity that I was often in a state of great agitation, the more so as I found it rather difficult to shuttle between the two.

Though certainly dutiful, Heisenberg never seemed completely at ease in the academic world during his youth. At ease or not, his career was astonishingly meteoric. He entered the university in October 1920. In the fall of 1921 he submitted his first paper for publication, and by June 1922 he was discussing quantum physics with Bohr on an equal basis. In 1923, after receiving his doctorate, he became Max Born's assistant in Göttingen, succeeding Pauli in that position. In 1924 he went to Copenhagen with a fellowship; after eight months there he returned to Göttingen. A month after that, in the early summer of 1925, Heisenberg, not yet twenty-four years old, became famous for his discovery of matrix mechanics.

Short, sturdy, blond, and athletic, Heisenberg was often described at the time as resembling a farm boy. Audiences could scarcely believe that this "farm boy" was giving lectures on the most advanced topics in physics, much less that he was the leader in the new physics revolution.

Recognition came swiftly. By the age of twenty-seven Heisenberg was a professor, Germany's youngest, at the University of Leipzig. Yet two years later, despite all this success, he still wrote to his mother in a surprisingly plaintive tone: "I remember the time when I myself was at my liveliest, you know, about ten years ago; that was also the most beautiful time of my life, such that my happiness also transferred itself to others."

Because of Heisenberg, Leipzig became a center of quantum theory, attracting young physicists from around the world, but their companionship seems to have done little to alleviate his loneliness, nor are those young physicists' recollections of Heisenberg marked by the kind of warmth they felt for Bohr, Ehrenfest, or Pauli. Most remember him as distant in personal matters, brilliant and intensely competitive, intolerant of losing at anything, even Ping-Pong.

Wolfgang Pauli

WHEN, AT EIGHTEEN, Werner Heisenberg began his studies in physics at the University of Munich, Arnold Sommerfeld assigned his most brilliant senior assistant, twenty-year-old Wolfgang Pauli, to grade Heisenberg's homework and to advise the youth on what courses to take. Pauli would soon be well established, thanks to his encyclopedic article on relativity theory, the one Einstein praised so lavishly. Having concluded for himself, with Sommerfeld's concurrence, that the exciting new frontier lay in understanding how quantum theory's rules applied to the atomic and molecular structure, Pauli easily convinced his young friend to follow his lead. Heisenberg now had a new father, Sommerfeld, and a new older brother, Pauli.

Heisenberg and Pauli remained linked in their intellectual interests from then until Pauli's death in 1958. The letters between them, starting in 1921, number in the hundreds; some are just brief notes, but many are lengthy, crammed with equations, providing an invaluable insight into the development of physics over almost four decades. Pauli in particular was a wonderful correspondent, ironic, often bitingly sarcastic, but with a wit that turned as easily inward as it did toward others. In turn, these letters often bring out the best in his correspondents. As Jorgen Kalckar, the editor of volume 6 of Bohr's *Collected Works*, observes,

> Bohr's scientific correspondence with Pauli is scientifically by far the richest and most illuminating—even compared to the Heisenberg correspondence which comes next to it in importance. Second, but even more decisive, is the vivid picture we receive from this correspondence of the warm human relationship between Bohr and Pauli. In general we miss in Bohr's written legacy the magic charm and spontaneity which made any conversation with him such an unforgettable experience. Indeed only in the correspondence with Pauli do we see him divest himself of the formality otherwise characteristic of the style of his writing. But here we catch at least a glimpse. We are allowed to share his delight in Pauli's witty sarcasms, which never really hide a feeling of deep friendship.

Pauli's letters to Ehrenfest are a case in point of his skill as a correspondent, and Ehrenfest's to Pauli are often a match. In their exchanges, instead of starting with *Dear Pauli*, Ehrenfest will sometimes begin *Lieber fürchterlicher Pauli* (Dear terrifying Pauli) or use a pet name, addressing his friend as *Sanct Pauli*, that being the name of the amusement and red-light district in Hamburg. He also gave his new young friend the epithet of *Geissel Gottes* (Scourge of God), one that Pauli was particularly proud of and used in signing his letters to Ehrenfest. Ehrenfest in turn signed his to Pauli as The Schoolmaster.

Ehrenfest and Pauli's's first meeting set the stage for their affectionate continuing Viennese *Witzkrieg* (war of wit). It took place during Bohr's 1922 Göttingen lectures, the Bohr Festspiele, the same gathering where

Heisenberg and Pauli first met the Dane. Ehrenfest had come to hear what his Danish friend had to say about quantum theory; encountering the arrogant young fellow Viennese turned out to be a side benefit. Ehrenfest's initial words to the author of the relativity review were reportedly, "Herr Pauli, your Encyclopedia article pleases me more than you do," to which Pauli is said to have replied, "How strange, my feelings toward you are just the opposite."

Pauli's collected correspondence also provides a window onto the changing relationships among physicists. In reading it, one sees him displaying a new informality with his senior mentors as he ages and gains recognition. The 1926 letters to Bohr begin *Sehr verehrter lieber Herr Professor* (Very honored dear Herr Professor) and sign off with a similarly flowery phrase. His 1928 letters start *Lieber Bohr* and end with *Empfehlungen an Deine Frau* (Regards to your wife). The more formal German pronoun address, *Sie,* has switched to the more intimate *du*.

More surprising, given their closeness in age, even Pauli and Heisenberg addressed each other as *Sie* until 1928. This was certainly due in part to the lingering reserve that, though diminished, still prevailed after World War I, but it may also be a reflection of a certain distance Heisenberg maintained even with his contemporaries. A great lifestyle difference also separated the two physicists. Heisenberg would have been happy to sleep on the beach while visiting the Bohrs at their country cottage in Tisvilde, but Pauli would have looked for the best hotel. The athletic Heisenberg, most comfortable in the woods and mountains, would rise early in the morning to work, while Pauli haunted coffeehouses and cabarets, going to sleep late and rising late. Corpulent and droopy lidded, his body characteristically swaying back and forth as he thought about a problem, Pauli smoked too much, drank too much, and ate too much. The abstemious Heisenberg refrained from any such excesses.

Though Pauli was very different from Heisenberg, he too encountered difficulties while making the transition from Wunderkind to established professor, chiefly for reasons that had nothing to do with physics. He was happy during most of a four-year stay in Hamburg, which followed his residency in Copenhagen and preceded his tenured professorship in Zurich. A jovial group of colleagues there was largely the reason. Arriving at age twenty-four in the historic northern German metropolis, Pauli quickly became something of a bon vivant. As he wrote to a friend in 1926, "I've found drinking wine very agreeable. After the second bottle of wine or champagne, I usually become good company (which I never am when sober) and can, under these circumstances, make a very good impression on my surroundings, especially if they are feminine!" However, family problems soon came to weigh heavily on him.

His father had fallen in love with a young woman and separated from Pauli's mother. In November 1927, distraught, she committed suicide by swallowing poison. Pauli was bereft. In a letter written many years after the events to the famous Zurich psychiatrist Carl Gustav Jung, Pauli said, "Alone in the express train to Zurich my mind went back to 1928 as I took the same route toward my new professorship and my great neurosis." It's hard to pin down what the "great neurosis" was, but its manifestations were clear.

In a new city and with new responsibilities, Pauli suddenly became prey to uncharacteristic impulsive actions. In May 1929 he abruptly left the Catholic Church. Six months later, just before Christmas, he married a dancer named Käthe Deppner. The marriage was a disaster. Two months after the wedding he wrote an old friend a long letter. It concludes with a postscript: "In case my wife runs away, you (just like all my other friends) will receive a printed notice."

Less than a year later, in November 1930, Käthe and Wolfgang were divorced. Before marrying Pauli, Käthe had met a chemist. Now she went back to him. Pauli reflected, "Had she taken a bullfighter, I would have understood, but an ordinary chemist . . ." This remark, in Pauli's characteristically acerbic way, captures not only his own insecurities but also the condescension many physicists at the time felt toward chemistry. Dirac had famously published a year earlier an article in the *Proceedings of the Royal Society* that said, "The underlying physical laws necessary for the mathematical theory of a large part of physics and the whole of chemistry are thus completely known." Even though Dirac had gone on to add that the calculations might be too difficult, the general mood among theoretical physicists was that the "whole of chemistry" had, in principle, been solved by quantum mechanics.

Despite his bravado, and obviously in pain, Pauli began drinking and smoking heavily. Worried by his son's behavior, Pauli's father suggested that he consult Jung. Although Jung supervised Pauli's analysis, he did not conduct the therapy personally, but when the analysis was over, this unlikely pair grew close. As the years went by, Pauli sent reports of more than a thousand of his dreams to Jung for examination. They published a book together called *The Interpretation of Nature and the Psyche*; their selected correspondence has also appeared in print under the suggestive title *Atom and Archetype*. However, Pauli seldom talked with his physics friends about these subjects. With them he was the same old critical, cynical Pauli.

Paul Dirac

WHEN IT COMES to extraordinary physicists, Heisenberg and Pauli seem almost in the range of normal compared with Paul Adrien Maurice Dirac. A famous mathematician, Mark Kac, once divided geniuses into two classes. He said there are the ordinary ones whose achievements might be emulated by intelligent people through enormous hard work and a good dose of luck. Then come the magicians, whose inventions are so astounding, so counter to all the intuitions of their colleagues, that it is hard to see how any human could have imagined them. Dirac was a magician.

Pauli and Heisenberg grew up in cultured households, the sons of university professors. They were molded by the senior figures of Sommerfeld in Munich, Born in Göttingen, and Bohr in Copenhagen, as well as by their exchanges with each other. Dirac's genius seems to have emerged full-blown with no help from the outside world.

Describing Dirac as reserved and taciturn is an understatement. There are reams of stories about him in fellow physicists' recollections, for example, how he responded only to precisely phrased questions, how an answer exactly addressed the question and nothing more. If he didn't answer at all, that wasn't Dirac being rude; the question had simply not been phrased so as to allow a precise answer. If someone would say during one of his lectures, "I don't understand how you derived that equation," Dirac would either ignore the speaker or reply, "That is not a question." Tall and thin, often slouching with hunched shoulders, Dirac in his stance seemed to underline his diffidence.

His Nobel Prize, awarded in 1933, was more a matter of apprehension than of celebration for the physicist. He was concerned about publicity and the invasion of his privacy, fearing the effects they might have on his reclusive life. He initially considered refusing the prize, but when Rutherford, his senior colleague at Cambridge, warned him that such a refusal would in fact generate more publicity, he reluctantly accepted Stockholm's recognition, making the trip to Sweden accompanied by his mother. A London newspaper at the time described him as being "as shy as a gazelle and as modest as a Victorian maid."

Dirac is a hero to theoretical physicists of my generation because of the beauty of his thoughts, still very much with us, the simplicity of his manner, and the directness of his speech. To give a flavor of the latter, here is one of the countless stories about his approach to life and to physics. In 1927 Dirac spent some time in Göttingen during which he became a friend of J. Robert Oppenheimer's, the two of them living in the same boardinghouse. One day, observing Oppenheimer reading Dante—in the original Italian of course—the ever logical and rational Dirac reportedly said to him, "How can you do both physics and poetry? In physics we try to explain in simple terms something that nobody knew before. In poetry it is the exact opposite." It is hard to guess what Oppenheimer must have thought of this remark. At one level he knew he didn't want to give up all his other interests, but he must have wondered if part of Dirac's success, already so manifest, was due to this single-minded dedication to physics.

I once described Dirac to my stepdaughter, who is both a neurologist and a psychiatrist, trying to illustrate for her how one can be a great theoretical physicist and yet, seemingly, behave oddly and lack other intellectual interests. She laughed and said that if Dirac were today to be examined in her clinic, he might be diagnosed as having "borderline pervasive developmental disorder." She then added that the test, however, would not reveal he was a true genius. But he was, one whose blending of mathematical elegance with physical reality was comparable to that of Einstein, whom he resembles, not in personality but in his quiet and solitary pursuit of the laws of physics. Bohr, who respected him deeply and could not help but admire his dedication, once said, "Of all physicists, Dirac has the purest soul," though even Bohr would occasionally shake his head at Dirac's literal interpretation of phrases.

It should come as no surprise that personality and scientific originality show little sign of being linked. Like any other profession, physics has talkers and quiet people, recluses and social butterflies. But genius, that rarest quality, may appear anywhere.

The theoretical physicist of the post–World War II generation whose work in many ways is most similar to Dirac's was Richard Feynman. In fact Feynman's brilliant derivation by so-called path integrals of a new formulation of quantum mechanics takes as a jumping-off point a 1932 paper by Dirac. The work on quantum electrodynamics for which Feynman, together with Julian Schwinger and Sin-Itiro Tomonaga, received the Nobel Prize is the modern version of a program started by Dirac in 1927. However, one cannot imagine two more different personalities: Feynman was the ultimate extrovert, and Dirac, the ultimate introvert. Yet, in the world of theoretical physics, that matters not at all. Each of them derived equations that, but for the passage of time, the other might have derived as well. In their depths, each understood, admired, and resembled the other. Their manners in life and their faces to the outside world were totally different; their styles in physics were similar. One of the great privileges of being a theoretical physicist is being able to see that inner resemblance.

While Feynman's character was molded in an outgoing Jewish household in Brooklyn, Dirac, born in 1902, grew up in a very controlled one in Bristol, England. His mother was English but his father was Swiss, a martinet who taught French at a local school. As Dirac remembered,

> My father made the rule that I should only talk to him in French. He thought it would be good for me to learn French in that way. Since I found that I couldn't express myself in French, it was better for me to stay silent than to talk in English. So I became very silent at that time—that started very early.

Dirac eventually learned grammatically correct French and was rewarded by having meals in the dining room with his father while his mother and siblings, failing the test, ate in the kitchen. However, he still did not speak.

Though he was socially awkward, his ability propelled him quickly through school, and he graduated from Bristol University with an engineering degree, just as his older brother had. This brother had wanted to become a physician but was forced by his father to study engineering, did badly, and committed suicide at age twenty-four. Paul, on the other hand, was a brilliant student, shining particularly in mathematics, and graduated at nineteen. He stayed for two years of graduate school at Bristol and then went to Cambridge with a scholarship he had won earlier.

At the time Cambridge was probably the world's greatest center of experimental work in atomic and nuclear physics. Rutherford had moved there from Manchester after World War I, bringing with him James Chadwick, the future discoverer of the neutron. Patrick Blackett, C. T. R. Wilson, Pyotr Kapitza, and John Cockcroft, all future Nobel Prize winners for experimental physics discoveries, were actively working there at the time in the company of past Nobel recipients Francis Aston and J. J. Thomson, but Cambridge's impact on theoretical physics was in no way comparable to that of Göttingen, Munich, or Copenhagen. Nevertheless it was a significant hub, and one could not help sensing there the growing excitement in quantum physics, fueled in part by the arrival of distinguished visitors. One of them was Bohr, who came to lecture in the spring of 1925 and of course to visit his old friend Rutherford. Dirac had this recollection:

> While I was very much impressed by him, his arguments were very much of a qualitative nature, and I was not able to really pinpoint the facts behind them. What I wanted was statements that could be expressed in terms of equations and Bohr's work very seldom provided such statements.

In July 1925, little more than a month after he had discovered the first form of quantum mechanics, Heisenberg gave a presentation at Cambridge. It is not clear whether Dirac attended the seminar or not; in any case it left no impression on him. However, in September Heisenberg sent the page proofs of his latest article to Ralph Fowler, Dirac's Cambridge adviser. A puzzled Fowler passed the paper on to Dirac. Now Dirac had equations in front of him. Two weeks later he knew what to do. He had discerned as the heart of Heisenberg's paper something that neither Heisenberg nor Pauli nor Bohr had quite recognized: the simple and essential difference between classical and quantum mechanics. He quickly wrote a paper about it.

Fame, certainly not sought by Dirac, came to him quickly. In Göttingen, Max Born and his student Pascual Jordan, who were in close contact with both Heisenberg and Pauli, had reached essentially the same conclusion as Dirac. They were, however, totally unaware of his existence, so when his paper reached them, they were stupefied. As Born said, "This was—I remember well—one of the greatest surprises of my scientific life. For the name Dirac was completely unknown to me, the author appeared to be a youngster, yet everything was perfect in its way and admirable."

Classical Mechanics versus Quantum Mechanics

THOUGH IT WAS NOT CLEAR at the start, appreciating the ways in which quantum mechanics differs from classical mechanics was a watershed for physicists in their understanding of natural phenomena on the minute scale of atomic lengths. It isn't easy for nonphysicists to appreciate the subtleties of the arguments on the subject, but it was also far from easy for physicists in the late 1920s and early 1930s to do so. Their own difficulties in coming to terms with the distinction between the classical and the quantum were lampooned in the *Faust* skit as a takeoff of Goethe's Walpurgisnacht scene, in which Faust absconds to the Harz Mountains with Mephistopheles to consort with witches. Walpurgisnacht, the night before May Day, was traditionally celebrated as the moment when spring finally triumphs over winter. Before that can happen, legend said, witches have to be allowed one last revel.

Delbrück/M.C. with the classical mechanics' Walpurgisnacht.

The Master of Ceremonies (M.C.), played by Delbrück himself, appeared for the first time in the skit at this point, presumably attempting to drive away the Classical and replace it by the Quantum, spring after winter. Faust asks him,

What's going on today?

to which Delbrück/M.C. replies,

Walpurgis Nights: the Classical Poetical,
And afterwards, the Quantum Theoretical

Faust argues with him about the difference between the two. Delbrück/M.C. tries to explain it to him:

Faust, you must expect
That with the Classical there's no effect
Upon the Audience.

But Delbrück, only a budding physicist, does not speak with the authority Faust is seeking. He therefore asks for some greater expert to speak before he will be convinced. At this point, Dirac's stage counterpart (he is simply "Dirac" in the script) appears, underlining Delbrück's assertion with his own proclamation:

Correct! Correct!

Faust tries arguing one more time, but Dirac weighs in with finality:

Not allowed!

Dirac had become the authority on the difference between classical and quantum mechanics by his interpretation and extension of what Heisenberg had said about the subject in his groundbreaking 1925 paper. I quoted from the paper earlier in attempting to explain why physics advances in the 1920s were mainly due to theoretical insights. Here is a fuller quote from the same source, in which Heisenberg expresses his opinion on the subject:

Dirac and Ehrenfest/Faust in discussion.

> Even for the simplest quantum-theoretical problems, the validity of classical mechanics simply cannot be maintained. In this situation it seems sensible to discard all hope of observing hitherto unobservable quantities, such as the position and period of the electron, and to concede that the partial agreement of the quantum rules with experience is more or less fortuitous. Instead it seems more reasonable to try to establish a theoretical quantum mechanics, analogous to classical mechanics, but in which only relations between observable quantities occur.

Heisenberg declares that "the validity of classical mechanics simply cannot be maintained." Physicists had to build up a new framework to replace it. He has taken the all-important first step, but he is unsure how to proceed. To understand the problem he faced, a small digression is necessary.

Mechanics deals with ropes, pulleys, inclined planes, rotations, stresses, and strains, but its essence is the study of how objects move or don't move when subject to forces. Though it has been perfected over the centuries and much of modern mathematics was developed to deal with problems that arose in studying its predictions, classical mechanics' basic laws were already stated by Newton at the end of the seventeenth century. The first major change to these laws took place in 1905 when Einstein reformulated our notions of space, time, and simultaneity. In a practical sense, however, relativity theory's corrections to classical mechanics' determination of macroscopic objects' motions are insignificantly small except when speeds begin to approach the ultimate limit, the speed of light. This means that Newton's formulation of mechanics is more than adequate to describe all our familiar objects, from falling apples to spinning planets.

However, during the twentieth century's first decades, as physics began to examine motion on the very small scales of atomic distances, a whole new set of questions arose. Did electrons move in orbits about the nucleus according to the same laws that cause apples to fall or planets to circle the sun? The forces were electric in one case and gravitational in the other, but the crux of the question was, Did Newton's laws of motion continue to hold true? In his 1913 model of the atom, Bohr had broken with classical theory by assuming that the radii of electron orbits were fixed by a set of conditions that related to quantum theory. He had also assumed that, contrary to what had been expected, electrons only emitted or absorbed radiation when they jumped from one possible orbit to another. His model's preliminary ageement with experiment was spectacularly good, but evidence in the following years, both from experiment and from further calculations, showed that the picture Bohr had proposed was inadequate. By 1925 Heisenberg and Pauli were maintaining that the whole view of electrons moving in orbits needed to be abandoned. But if that was the case, what were the new laws that dictated motion on the atomic scale? What new mechanics was going to replace classical mechanics as a tool for describing atoms? Maybe finding the answer would require a leap into the unknown.

Heisenberg took the first step by formulating what he called matrix mechanics, a mechanics based on treating only quantities that could be measured in experiments. This did not include electron orbits, deemed by him to be not measurable or, as he said, unobservable.

Dirac's paper, written shortly after studying Heisenberg's, recognized that the necessary transition away from classical mechanics was simpler than Heisenberg had supposed. Dirac maintained that

> Only one basic assumption of classical theory is false . . . the laws of classical mechanics must be generalized when applied to atomic systems, the generalization being that the commutative law of multiplication, as applied to dynamical variables, is to be replaced by certain quantum conditions.

Position (q) *and momentum* (p), *differently understood in quantum mechanics than in classical mechanics.*

These conditions were equivalent to stating that the product of an electron's position times its velocity is not the same as the product of velocity multiplied by position: A times *B* does not equal *B* times A. The difference is instead proportional to Planck's constant, the quantity introduced in our earlier discussion of quantum theory's birth.

This replacement, the heart of quantum mechanics, is no small matter. Heisenberg's assertion that "the validity of classical mechanics simply cannot be maintained" is too strong to be taken literally, but Dirac's "Only one basic assumption of classical theory is false" understates the magnitude of the necessary change. In practical terms, because the techniques of classical mechanics were familiar, at least physicists now had a good idea of where to begin. The electron orbits might not be observable, but at least physicists had a new set of rules to replace those dictated by Newton's laws. They could attempt to calculate the predictions of radiation emitted or absorbed by moving electrons even if they couldn't describe the details of the motion that resulted in that emission. Clear progress was being made toward developing what came to be known as quantum mechanics.

This still did not solve the whole problem: even with the necessary formalism for computations in hand, it wasn't clear what the answers to those calculations meant. Their interpretation would take two more years, culminating in the struggle and final accord between Bohr and Heisenberg: their development of the principles of complementarity and uncertainty that form the basis of the Copenhagen interpretation of quantum mechanics.

Meanwhile, doctorate in hand, Dirac left England in late 1926 for the Continent. Where to? Well of course, to Copenhagen. Dirac remembered his four-month stay there warmly: "I admired Bohr very much. We had long talks together, long talks during which Bohr did practically all the talking." Continuing from Copenhagen on his tour of major theoretical physics research centers, Dirac went to Max Born's Göttingen for a sojourn and then on to Paul Ehrenfest's Leiden. In October 1927 he arrived in Brussels for the fifth Solvay conference.

These important conferences gathered together every few years a small number of leading researchers in a predetermined but varying field of physics, but unlike the Copenhagen meetings, they did not emphasize either informality or youth. However, one noticeable feature of the 1927 event was that three attendees were men in their midtwenties, the youngest trio ever invited to a Solvay conference. Born in 1900, 1901, and 1902, they were of course Pauli, Heisenberg, and Dirac. No longer revolutionaries, they were already part of the establishment.

In the summer of 1929, Dirac and Heisenberg, now ensconced as two young pillars of the establishment, went on lecture tours of the United States. Meeting in the Midwest, they journeyed together to California and then by steamer to Japan, a trip arranged for them by the physicist Yoshio Nishina, a friend from their times in Copenhagen. The memoirs of the two travelers contain several amusing stories. These include vintage anecdotes about each, told by the other. First, this is from Heisenberg's story, about a dance aboard ship:

> We were on the steamer from America to Japan, and I liked to take part in the social life on the steamer and, so, for instance, I took part in the dances in the evening. Paul, somehow, didn't like that too much but he would sit in a chair and look at the dances. Once I came back from a dance and took the chair beside him and he asked me, "Heisenberg, why do you dance?" I said, "Well, when there are nice girls, it is a pleasure to dance." He thought for a long time about it, and after about five minutes he said, "Heisenberg, how do you know beforehand that the girls are nice?"

Now Dirac's, about ascending a high tower in Japan that had a platform ringed by a stone railing:

> Heisenberg climbed up on the balustrade and then on the stonework at one of the corners and stood there, entirely unsupported, standing on about six inches square of stonework. Quite undisturbed by the great height, he just surveyed all the scenery around him. I couldn't help feeling anxious. If a wind had come along then, it might have had a tragic result.

Heisenberg's story is one more in the collection of amusing tales about the socially awkward Dirac scanning sentences for their logical construction. Dirac's anecdote is richer in its allusion, though he certainly only meant to provide a literal account. Heisenberg, known in his youth as the greatest risk taker of his Neupfadfinder group, came to be regarded as the most daring physicist of his generation.

They separately returned to Europe from Japan, Dirac via Siberia and Heisenberg via India. In December 1933 they were together again, this time in Stockholm to receive their Nobel Prizes. (Heisenberg's 1932 prize was awarded in 1933.) They were applauded for their brilliance, and since they were both still so young, more great things were expected of them. But would they and Pauli be able to sustain the momentum into their thirties and forties?

The new generation, while deeply respectful of the trio's achievements, had concerns of its own, not the least of which was making their own contributions, hopefully before they in turn reached thirty. The under-thirties in Copenhagen were a very distinguished group. Delbrück, Bloch, Subrahmanyan Chandrasekhar, Lev Landau, Nevill Mott, and others went on to win Nobel Prizes; some, like Nishina from Japan and Homi Bhabha from India, spread the doctrine in faraway lands. Still others, such as Casimir, Peierls, and Weisskopf, while not winning Nobel Prizes, became major figures in the world of physics. But none of them quite achieved the status of Dirac, Heisenberg, and Pauli, or that of Bohr. There are many reasons for this, but one of them is surely the fact that no matter how interesting and how important physics became, the revolution was by and large over. It was no longer the time when, as one of the 1930s Copenhageners put it, "Every Ph.D thesis opened up a new field of applications for quantum mechanics."

Chapter 6

The Front Row: The Young Ones

STUDENT
But recently I've quitted home,
Full of devotion am I come
A man to know and hear, whose name
With reverence is known to fame.

Goethe, *Faust, Part I,* 1513–16

The Curse of the *Knabenphysik*

HEISENBERG, PAULI, AND DIRAC also created for theoretical physics what I like to call the curse of the *Knabenphysik,* the notion that one should have done something of significance before turning thirty. It is not an entirely new phenomenon; after all, Newton was only twenty-four when he formulated the theory of gravitation *and* developed the basic notions of calculus. But the legend of the curse really began to take hold in 1905 when twenty-six-year-old Einstein laid the foundations of quantum theory *and* presented his special theory of relativity. It grew when Bohr explained, at age twenty-seven, the motion of electrons around the atomic nucleus, and it was reinforced by the arrival on the scene of the *Knaben* trio: Heisenberg, Pauli, and Dirac.

I don't know why these gifts manifest themselves so early. One would think that knowing more and growing wiser would be a bonus, as it seems to be in the humanities. Perhaps it's a question of daring, energy, and single-mindedness, of having a totally fresh approach. In any case the curse of the *Knabenphysik* seems to be alive and well. Don't be misled by the gray hair of current Nobelists. Three theorists, David Gross, H. David Politzer, and Frank Wilczek, shared the 2004 physics prize, but it was awarded for work they had done thirty years earlier. Gross, the old man of the trio, had then been thirty; Politzer and Wilczek were under twenty-five.

There are no senior tournaments for theoretical physicists, unlike those for athletes, but all is not lost. Many satisfactions remain, including that of simply continuing to play the game. It's also a pleasure to see new talent in physics emerge, even if it does perpetuate the curse.

Michael Cohen, a retired colleague of mine, was recently reminiscing with me about his own early days in research, almost fifty years ago. At the time he was a very successful graduate student of Richard Feynman's. They worked together and published a few papers jointly. He told me the following story. While sitting one day in a Caltech seminar room waiting for a talk to begin, Cohen found himself wondering what happens to a physicist as he ages. He turned to Feynman, who was next to him, and asked his mentor, "How will I know when physics passes me by?" Feynman simply smiled. It was a beautiful spring day and the windows were open, a gentle breeze blowing outside. Cohen, who had been working late the night before, dozed off. He next felt a nudge in the ribs. Waking with a start, he heard Feynman say to him, while pointing to the open window, "There it goes, Mike. Kiss it good-bye."

Max Delbrück

AT THE TIME Cohen was kissing it good-bye, one of Feynman's most distinguished colleagues and best friends at the California Institute of Technology was the same man who had sat a quarter century earlier between Paul Ehrenfest and Lise Meitner in the front row of the Copenhagen group picture. Max Delbrück had then been only twenty-five years old.

Though his fame rests on his place as one of the founding fathers of molecular biology, not on his work as a physicist, he was in many ways Bohr's closest spiritual heir, the one among the young in Copenhagen who tried hardest to keep alive Bohr's quest for the meaning of complementarity, the notion that Bohr pioneered in his study of quantum mechanics and later tried to extend to other fields of knowledge. Not surprisingly, Delbrück regarded Bohr as his most important mentor. Furthermore he always thought of himself as a physicist, even after he had de facto become a well-known biologist. In 1949 he summarized his views in an essay entitled "A Physicist Looks at Biology," and in 1969, even as he received the Nobel Prize in Physiology or Medicine, he chose as the topic for his acceptance speech "A Physicist's Renewed Look at Biology: Twenty Years Later."

Tall, lean, and boyish-looking, Max Delbrück was the seventh and youngest child of Hans and Lina Delbrück. His was a distinguished family on both sides, generations of academics, lawyers, and civil servants. Max's father was once a member of the Prussian Parliament and later a professor of German history at the University of Berlin. The family residence was in the lovely Grunewald area of Berlin, developed toward the end of the nineteenth century; it was a milieu full of German culture: music, poetry, art, and science. The Plancks were friends and neighbors. Families gathered for picnics, small piano recitals, evenings of lieder, discussions for the adults, and games for the children.

At first intending to become an astronomer, Max Delbrück changed his mind after hearing about the growing excitement in atomic physics. He thought the specific turning point for this decision had been a talk given in 1926 at the University of Berlin. Only nineteen, he was in the lecture hall when he overheard Einstein say to a colleague that the day's presentation was going to be about a very important piece of work. The subject was matrix mechanics, and the speaker that day was Werner Heisenberg, twenty-four years old. Delbrück didn't understand much that was said, but sensing the excitement, he vowed to become part of the new movement, to try to follow in Heisenberg's footsteps. That summer he went to Göttingen to study quantum theory.

Four years later Delbrück won one of the coveted Rockefeller Foundation fellowships. Set up in the 1920s, they provided a year's stipend at a site of the recipient's choosing. Given the general high quality of the candidates selected and the freedom that came with the awards, the fellowships played an important role in the physics of that time, allowing the young to move from one place to another, one country to another, one inspiring teacher to another. Splitting his fellowship year, as most theoretical physicists did, Delbrück chose to spend the first half with Bohr in Copenhagen and the second with Pauli in Zurich, the two places he felt he would learn the most. Both stays proved extremely important for his intellectual and personal development, each marking the beginning of a lifelong friendship as well as a crucial stage in his education.

Bohr and Pauli independently recognized special gifts in Delbrück, but they were also aware that theoretical physics might not be the best avenue for those talents. Since he had good insights and imagination, but lacked mathematical finesse, they gently steered him in other directions. Perhaps sensing that Delbrück needed support, not criticism from the Scourge of God, Pauli seems to have been uncharacteristically gentle in his attitude toward him. In any case the warmth between the two of them endured.

Bohr was more directly influential. With his encouragement and Pauli's tacit assent, Delbrück slowly moved toward biology. Steeped in atomic physics, he began to ask if there might be some analogy between genes, the units of heredity, and atoms, the units of matter. Could this analogy be extended to mutations and quantum jumps? How and where could he pursue these ideas?

In 1927 Hermann Muller announced that he had succeeded in inducing mutations in fruit flies by irradiating them with X-rays, a suggestive hint that the analogy between mutations and quantum jumps might not be so far-fetched. It just so happened that in 1932 Muller was moving to Berlin to work with the Russian émigré Nicolai Timoféef-Ressovsky, the head of a genetics research branch of the Kaiser Wilhelm Institute; it also seemed likely that Muller and others in Timoféef-Ressovsky's group had interests that might be similar to Delbrück's.

However, Delbrück needed an academic position that would allow him the freedom to explore his growing interest in biology. Lise Meitner provided him with the means for his metamorphosis by offering him a five-year position in her laboratory. As he wrote to Bohr in June 1932, "I have accepted Lise Meitner's offer to go to Dahlem as her 'family-theorist' in October largely because of the neighborhood of the very fine Kaiser Wilhelm Institute for Biology, to which I am entertaining friendly relations."

Delbrück was a prime example of how Meitner, an experimentalist, worked closely with theoretical physicists. Having a talented young one in her laboratory in Dahlem (a suburb of Berlin) proved useful for her as well as for Delbrück. There was another reason for his move back to Berlin. He could live at home with his mother, helping her overcome her loneliness after the death in 1929 of his father. The large house was also perfect for an informal series of seminars and discussion groups that he organized in which he brought together people from all different fields to discuss problems in genetics. It was all very un-German, very much not in the Herr Professor mold, but it worked. It was more in the spirit of Copenhagen.

Chapter 7

The Coming Storm

WAGNER
Your pardon! 'tis delightful to transport
Oneself into the spirit of the past,
To see in times before us how a wise man thought,
And what a glorious height we have achieved at last.

Goethe, *Faust, Part I,* 222–25

The Periodic Table

THUS DID OUR seven main characters approach the Miracle Year of 1932. It is time now to pick up the story of quantum theory's development and the passions aroused by the experts' differences of opinion.

Bohr's 1913 model of the atom had quickly made the physics and the chemistry communities focus on the great physical science classification system, the periodic table of elements. There may be no better way to explore what this arrangement means than to see it through the eyes of an exceptional child with a growing interest in science.

The well-known neuroscientist Oliver Sacks has written several books about patients suffering from neurological oddities, usually emphasizing the human aspects of their strange conditions. In a memoir entitled *Uncle Tungsten,* he instead turned to his own past, describing vividly his early delight with chemistry. Reminiscing about those childhood days, he recounts a 1945 trip he took to visit the newly reopened

London Science Museum. Looking up as he entered the building, the twelve-year-old Sacks saw some ninety open cubicles of dark wood arranged one next to the other on the wall at the head of the stairs. It was a giant display of the periodic table of elements, each cubicle labeled with a chemical symbol and containing, when possible, a sample of the element. As a fledgling chemist, he was already familiar with many of them, but seeing them all arranged before him was a revelation:

> To have perceived an *overall* organization, a superarching principle uniting and relating *all* the elements, had a quality of the miraculous, of genius. And this gave me, for the first time, a sense of the transcendent power of the human mind, and the fact that it might be equipped to discover or decipher the deepest secrets of nature, to read the mind of God.

The notion that all matter, living or nonliving, is made up of endless combinations of these elements is extraordinarily powerful. Combining it with the concept that the elements can be grouped into connected families forms the central dogma of atomic structure and the beginning of our modern understanding of relationships in chemistry. History classrooms around the world have various maps displayed on their walls, but all physics and chemistry classrooms prominently display this same large periodic table, their universal map.

Beyond the beautiful classification lies a question young Sacks did not ask himself that day in 1945. What leads to a *superarching principle uniting and relating all the elements?* This is what excited the young physicists in the 1920s, that brought them to Copenhagen, that they triumphantly answered, leading to Dirac's bold end-of-the-decade statement: "The underlying physical laws necessary for the mathematical theory of a large part of physics and the whole of chemistry are thus completely known."

The journey that led to their finding the explanation for the periodic table, its superarching principle, began with Rutherford's picture of the atom, a tiny nucleus carrying positive electric charge surrounded by electrons, each carrying a negative electric charge, symbolized as $-e$. The atom's known overall electric neutrality was guaranteed by having Z electrons surrounding a central nucleus whose charge was $+Ze$: $-Ze$ and $+Ze$ neatly balanced each other out.

This new image of the atom changed one of the basic assumptions that had shaped the periodic table. Before then each element had been labeled by the weight of its representative atom. Taking hydrogen's mass as one unit, helium's mass was four, lithium's seven, beryllium's nine, boron's eleven, carbon's twelve, and so on. In 1869 Dmitri Mendeleev, a bearded visionary Russian from Siberia, using these weights and observed chemical properties, introduced a grouping of elements into families. His work, ignored at first, dramatically prophesied the existence of three new elements. In the following twenty years, the significance of his grouping became increasingly evident as, one by one, each of the three was found, all with the properties and atomic weights Mendeleev had predicted.

But more accurate measurements showed that atomic weights were only approximately equal to integers like 4, 7, and 8, suggesting there might be a deeper rule that determined the elements' order. With Rutherford's picture of the atom, the analysis changed; the number of electrons an atom held became the key to labeling and to understanding the chemical behavior of an element. In this new arrangement the steps from hydrogen to carbon are a simple sequence of integers: 1, 2, 3, 4, 5, 6. One can continue all the way to uranium, the final stable element; it has ninety-two negatively charged electrons circling a nucleus carrying an equal positive charge. The classification was now neat, but the superarching principle was still missing.

In 1913 Bohr, having learned the details of Rutherford's model of the atom during his recent Manchester stay, took the next step by asking himself if quantum theory could determine the possible paths of the electrons circling the atomic nucleus. Might this be the key to the superarching principle? Because Rutherford's atom looked like a miniature solar system, proceeding by analogy made sense.

There had been a long and distinguished history of attempts to set the size of the planetary orbits around the sun through a simple mathematical formula, preferably one that would shed light on a deeper unrecognized truth of nature. If the letter A represents the average distance between Mercury and the sun, we now know that the respective distances between the next four planets and what Pythagoras called the central fire are 1.86A (Venus), 2.58A (Earth), 3.94A (Mars), and 13.4A (Jupiter). Many thinkers had asked if there was some explanation for these distances, or rather, for the different and often tidier values they thought them to be. In the end all their tries had failed. Those numbers are just what they are—1.86, 2.58, 3.94, and 13.4. There is no explanation for them.

Bohr knew, however, that failure in the solar system did not necessarily mean he would fail in the analogous case of the atom. If the lowest electron orbit had a radius we'll call *B*, might the successive orbits have radii 2*B*, 3*B*, 4*B*, 5*B*, and so on? Could those numbers 2, 3, 4, and 5 be linked to quantum theory? Could this be the beginning of a new chapter of science?

Pythagoras, the great sixth-century-BCE philosopher, was the first to think that there might be an explanation for the sizes of the planetary orbits. This early genius had founded a school with the credo "All things are numbers." His belief in their paramount importance, initially based on geometric considerations, had been strengthened by the discovery that the sounds of simultaneous vibrating strings are only harmonious if the string lengths are fixed as simple numerical ratios. Linking numbers and music was the Pythagoreans' first major achievement.

Attempting to extend their number-guided view of the universe, they next postulated that the earth and the planets were moving spheres, suspended in space, each completing a circular orbit about a central fire. The Pythagoreans envisioned every planet emitting a musical note whose frequency was set by its distance from the central fire. The resulting ensemble, assumed harmonious, could only be so if the ratios of the planetary orbit sizes were simple numbers, just as plucked strings' notes only blended when their lengths were ratios of simple numbers. In a leap joining numbers, music, and astronomy, they had created the concept of the harmony of the spheres.

Johannes Kepler, born two thousand years after Pythagoras's brotherhood in Croton, in southern Italy, had disbanded, kept alive the notion that some mathematical relation fixed the distances of the planets from the sun. His first youthful effort at planetary placement was a geometric one, based on inscribing spheres within regular geometrical solids, the whole founded on the recently promulgated Copernican model of planets circling the central body. But a meticulous study of data forced Kepler to accept that planetary orbits are ellipses, not circles. Searching over the next decades for an accurate description of those elliptical orbits, Kepler formulated the three laws of planetary motion that have made him immortal. But, try as he might, he could not easily explain why his three laws held. Continuing to grope for some justification, Kepler fell back on a scheme Pythagoras might have adopted: the ratios of the major and minor axes of the ellipses were set by musical harmonics.

This is one of those great transitional moments in the history of science: data, not some preconceived notion, was Kepler's guide in finding his three laws. This is why they remained true, even after Newton's majestic insights. Kepler's observations were correct, his explanation of them obsolete. And yet the collection of ideas that drove him onward remains alive and continues to inspire.

Years after the quantum mechanics revolution, Pauli tried to understand what had led Kepler to his results, hoping also to understand the quantum theory revolution Sommerfeld had introduced him to. In a special volume prepared for his teacher's eightieth birthday, Pauli wrote,

> It is as though there was here an echo of Kepler's search for the harmonies in the cosmos, guided by the musical feeling for the beauty of just proportion in the sense of Pythagorean philosophy—his geometry is the archetype of the beauty of the universe. And how admirably did Sommerfeld understand the art of communicating to his large circle of pupils his infallible feeling for the just proportion and for the harmonious.

In later years, trying to find the link between the inspirations and connections in his own life, Pauli attempted to match Jung's "primordial images or archetypes" with Kepler's view of images implanted in the soul (Kepler called them *archetypalis*):

> The process of understanding nature as well as the happiness that man feels in understanding, that is, in the unconscious realization of new knowledge, seems thus to be based on a correspondence, a matching of inner images pre-existent in the human psyche with external objects and their behavior.

Another way of putting it might be to ask, Does the outside world correspond to pictures already formed in the mind, or do these external objects shape how we organize the pictures? Bohr would probably have maintained that they are complementary points of view, both true.

There is something magical about Kepler, a link between the very modern and the distant past. Max Delbrück chose this historic scientist's life as the topic of his high school valedictory address. Preparing himself for speaking to his fellow students, teachers, and parents, he was allowed to consult rare books in the Berlin University library. In his own words,

> To see and handle these old books, 300 years old, was a tremendous experience, especially when I found in one of them some of the

speculations about the celestial harmonies expressed in terms of musical notes. It showed that Kepler literally thought in terms of a celestial dingdong at that time, rather than in terms of abstract mathematics.

Whether to be guided by mathematics or by a celestial dingdong, as Delbrück colorfully put it, would turn out to be a running argument between Bohr and Heisenberg in 1926, a source of much tension between them as they struggled to understand quantum mechanics.

The choice is a matter of preference, of style, but most theoretical work in science is a mixture of intuition, study of data, and mathematical analysis. All approaches are valid; sometimes one mechanism succeeds, sometimes the other, and sometimes it is not clear how a scientist has reached his or her conclusions.

But this parenthesis on Kepler has taken us away from the thread of our story of how Bohr discovered his model of the atom, and how that discovery eventually provided the superarching principle that explains the periodic table of elements.

The New Kepler

AN OLD FRIEND of Bohr remembered him in 1913 as rushing, young and impatient. "He was an incessant worker and seemed to be always in a hurry. Serenity and pipe smoking came much later." Just back from England, Bohr was trying to establish himself in Copenhagen. What better way to do that than by following up on his interest in atomic physics, already piqued by his stay in Rutherford's Manchester? He began by concentrating on hydrogen, the simplest of all elements, the one whose atom has only a single electron.

Applying quantum principles, Bohr assumed the hydrogen atom radiates energy in the form of quanta, as Planck and Einstein had shown, and furthermore that a quantum's energy was simply Planck's constant multiplied by the quantum's frequency. By overall conservation of energy, each individual quantum's energy had to match the difference of the atom's energy before and after it emitted the quantum, so Bohr began thinking that the best place to search for a clue to an atom's possible energies would be in its patterns of radiation. He found what he was looking for in early 1913, when a friend informed him of a surprising relation between the different frequencies of hydrogen's so-called spectral lines.

For over fifty years scientists had known that chemical compounds, when sufficiently heated, emit radiation of many different frequencies. Passing such a multifrequency beam of radiation through a prism spreads it out into a wide display, each frequency appearing as a separate spectral line. The frequencies of these lines differ from compound to compound, from element to element, the pattern of lines so clear and so distinct that by the end of the nineteenth century their analysis had become a powerful chemical diagnostic tool. Carefully tabulated, each line's frequency was believed to be significant only as a label of the chemical substance being studied.

Some probing observers, not necessarily believing that anything notable would emerge, began to look for connections between the frequency values. In 1885 a sixty-year-old Swiss secondary-school teacher named Johann Balmer found that hydrogen atom spectral frequencies were proportional to a constant, the same for all lines, multiplied by the difference of inverse squares of integers, for example, $1/2^2 - 1/3^2$, which is $1/4 - 1/9$. This knowledge was filed at the time as an interesting piece of information. Almost thirty years later the curio turned out to be Bohr's crucial first insight into the quantum theory of the atom.

Balmer's formula led Bohr to a construction of a hydrogen atom model, with electron orbit radii and therefore electron energy values fixed by the very same integers the Swiss schoolteacher had introduced. They were now called quantum numbers.

In 1962, shortly before he died, Bohr gave an interview to a distinguished historian of science in which he reminisced about what the

Balmer formula had meant to him personally and to quantum theory's development:

> One thought spectra are marvelous, but it is not possible to make progress there. Just as if you have the wing of a butterfly, then certainly it is very regular with the colors and so on, but nobody thought one could get the basis of biology from the coloring of the wing of a butterfly. So that was the way to look at it.

And nobody had thought one could see the foundations of atomic theory from the frequencies of spectral lines.

Bohr's model was extraordinarily interesting, but did it agree with experiment just by coincidence? Doubts were quickly laid to rest by a clear and dramatic prediction he made regarding helium, number 2 in the periodic table. Helium is rare on earth but abundant in the sun, so much so that it was first observed in solar spectral lines and thereby acquired its name from *helios*, sun in Greek. A normal helium atom has two electrons, but if it is sufficiently heated, one of those electrons will escape, leaving an ionized helium atom. The remaining helium electron can jump from one allowed orbit to another, looking very much like the single electron in hydrogen, but according to Bohr's model the corresponding spectral frequency formula should differ by a factor of four from hydrogen's because of the difference in electric charge between the hydrogen and helium nuclei.

The high temperatures of that solar environment suggested that spectral lines from ionized helium should be visible. They were, and they differed from hydrogen's by the Bohr-anticipated factor of four, a remarkable confirmation of his theory. However, very soon thereafter a British experimental physicist observed ionized helium lines in a laboratory experiment, obtaining a more accurate measurement than the one achieved through astronomical determinations. His answer was 4.0016. He concluded that Bohr's model was wrong because its prediction lay outside the very precise experiment's tiny margin of error.

Bohr quickly responded that he had, for simplicity's sake, made the approximation that the electron's mass was negligibly small compared with that of the nucleus. Inserting the known mass values in his formula, his more accurate prediction was 4.00163. Agreement of this kind between theory and experiment had never been reached in atomic calculations. It was a sensation! The news spread quickly that a young Dane had found something very, very significant.

Georg von Hevesy, Bohr's good friend from their days together in Manchester, heard about the results. Working at the time in Germany, Hevesy quickly went to see Einstein. He reported back to Bohr that Einstein had been noncommittal at first, but when told about the helium result, his eyes lit up and he said to him, "This is an enormous achievement. The theory of Bohr must then be right." The agreement with experiment of the predictions was inescapable, so striking that at least some aspects of the new atomic quantum theory had to be correct.

Like Kepler, Bohr had great intuitive powers. In formulating his theory of the atom and again, in deciphering the periodic table, he made assumptions that, like Kepler's, were prima facie wrong, and yet he often obtained a correct final answer, sometimes baffling those attempting to follow his reasoning. Ehrenfest, skeptical of this reliance on intuition, wrote Lorentz in 1913, "Bohr's work on the quantum theory of the Balmer formula has driven me to despair." But he was convinced of its importance; his own adiabatic theorem, developed for other purposes, quickly became one of the chief tools in studying the theory.

Like Ehrenfest, Arnold Sommerfeld found his excitement growing as he studied Bohr's work. He too felt there had to be some deep truth in it. By now in his late forties, this short, gruff man with a bristly mustache believed, correctly as it turned out, that his own mathematical acuteness might be valuable in making further progress. Just as Kepler had generalized Copernicus's circular paths to ellipses, he imagined a similar extension of Bohr's: in his hands the single integer that had labeled the possible radii of the electron's circular track now became three integers specifying an ellipse's size, its eccentricity, and its orientation in space.

Furthermore, any atom more complicated than hydrogen possessed necessarily a greater number of electrons, and therefore a researcher had to specify multiple electron orbits. This required the selection of more integers, or quantum numbers, than three. Choosing them judiciously required a healthy dose of faith in the underlying theory. It wasn't always easy, but Bohr seemed to instinctively know how to do it, leading one prominent theoretical physicist to quip, "In Copenhagen they could quantize your grandmother."

Sommerfeld's book *Atomic Structure and Spectral Lines*, first published in 1919, quickly became the bible of the emerging theory. Its preface linked Bohr's theory to that of Kepler and Pythagoras:

> What we are hearing nowadays of the language of spectra is a true music of the spheres within the atom, chords of integral relationships, an order and harmony that becomes ever more perfect in spite of the manifold variety. The theory of spectral lines will bear the name of Bohr for all time.

He communicated his own sense of awe to his students, most particularly to Pauli and Heisenberg, the two new ones he was acquiring as the book was being published.

As these students learned the tools of the trade, it was once again a time of hope in physics and in the world. The war was finally over, and the clouds shrouding the atom's mystery seemed to be lifting. The excitement in physics generated by Bohr's breakthrough had progressed more slowly than it would have otherwise because World War I started soon after the appearance of his first papers on atomic theory. Now most of the theory's architects were eager to begin conversing with one another again.

All but one of our seven front-row characters escaped service in the war. Ehrenfest and Bohr were citizens of neutral countries—the Netherlands and Denmark, respectively. Delbrück was still a child, Heisenberg would have been called up if the war had gone on a little longer, and Dirac was certainly too young. Pauli, barely old enough for the draft at war's end, was deferred because of *Herzschwäche* (syncope, or fainting from insufficient blood flow to the brain). Meitner, the only one of the group to serve during the conflict, spent a full heart-wrenching year as a nurse, first at the eastern front and later at the Italian one.

In retrospect it seems hard to understand how physics could have flourished in post–World War I Germany, given the general devastation of the country and the economic ruin that followed. Periodicals were scarce, and books were too expensive to buy. Much of the credit for the success must go to Planck and a few colleagues who created an organization known as the Emergency Association of German Science. It collected funds from government, from industry, and from abroad and then disbursed them as grants. These were awarded solely on the merit of the science and the promise of the individual, ensuring that the likes of Sommerfeld, Born, and Einstein were supported and their students not driven away from basic research.

The war years had been terrible for Planck, marked by the death of his son at the front and both his twin daughters from childbirth complications. Others grieved with him and for him, but they also realized that Planck's devotion to science and to duty was a balm for him. He wrote in a 1919 newspaper article, "As long as German science can continue in the old way, it is unthinkable that Germany will be driven from the ranks of civilized nations." He felt it was his duty to make sure that did not happen.

Göttingen in 1922

In the early 1920s, the mantle of leadership in quantum theory and atomic structure rested on Niels Bohr's broad shoulders. After his great success in explaining the hydrogen atom, he had begun to think about generalizing his techniques to the even harder task of explaining the periodic table of elements. It took him several years to mount a full scale attack, years in large part devoted to establishing himself in Copenhagen, assembling the financing for his new institute, and supervising its construction, but by late 1920 he was ready to present his ideas. In 1921 he published two short papers in *Nature* containing their key notions.

Bohr's conclusions were both intriguing and baffling. Five years earlier, Ehrenfest had written Sommerfeld: "Even though I consider it horrible that this success will help the preliminary but still completely monstrous Bohr model on to new triumphs, I nevertheless wish physics at Munich further success along this path." By 1921 Ehrenfest, now a good friend of Bohr's and a supporter of his ideas, found himself once again perplexed. In September he wrote the Dane: "I have read your *Nature* letter with eager interest . . . Of course I am now even more interested to know how you saw it all." Bohr's correspondence principle, the way he connected quantum and classical notions, puzzled Ehrenfest. Bohr's papers had pictures of intersecting elliptical electron orbits, charts, heuristic notions, but what mathematical calculations justified the construction? Was Bohr being deliberately obscure?

Many others beside Ehrenfest, eager to know how Bohr had reached his conclusions, were looking forward to attending a much-anticipated series of seven lectures at Göttingen that Bohr was scheduled to deliver in June 1921. If they missed those, there would be another chance a few months later to hear him at the fourth Solvay conference, held as usual in Brussels. Its chosen topic was "Atoms and Electrons," a perfect forum for Bohr to expound his theory.

Unfortunately Bohr was exhausted by the intense effort of simultaneously setting up the institute and doing research. At his doctor's recommendation he canceled all talks for the next few months, trying to rest, and therefore missed the 1921 Solvay events. Six months later, back at full strength, Bohr published a sixty-four-page article entitled "The Structure of the Atoms and the Physical and Chemical Properties of the Elements." Appearing originally in German, it was quickly translated into French and English and read eagerly by atomic physicists hoping to finally see the equations and the details of the algebraic calculations that explained the correspondence principle, but once again they were missing. It turns out there was a good reason for that. They didn't exist.

Hendrik Kramers, Bohr's first Copenhagen disciple and the physicist closest to him in those years, remembered what it was like:

> It is interesting to recollect how many physicists abroad thought, at the time of the appearance of Bohr's theory of the periodic system, that it was extensively supported by unpublished calculations which dealt in detail with the structure of the individual atoms, whereas the truth was, in fact, that Bohr had created and elaborated with a divine glance a synthesis between results of a spectroscopical nature and of a chemical nature.

Kramers said the correspondence principle was "a somewhat mystical magic wand, which did not act outside Copenhagen." Only Bohr and he, to a lesser extent, knew how to apply it.

Bohr's success with hydrogen had been stunning, a bolt out of the blue from an unknown young Dane; his work since 1913 on atomic physics and his recent attempts to explain the periodic table marked him as the leading theoretical physicist treating such problems. His approach was new, seemingly based on examining a wide range of data, using his intuition as a guide toward a synthesis and then looking for an explanation of the whole. In Bohr's hands the method was yielding startling results. The anticipation to hear him and have a chance to question his findings was palpable.

In June 1922 the wait was over. Bohr went to Göttingen to give the seven lectures originally scheduled for the year before. It was going to be a two-week event with ample time left for discussions. Despite the difficulties in travel caused by postwar poverty, quantum theorists from across the continent made the trip. Ehrenfest came from Leiden; Sommerfeld traveled from Munich, bringing along Werner Heisenberg. With inflation still raging, Sommerfeld paid for his student's ticket and found him a couch to sleep on in a friend's house. The new wunderkind Wolfgang

Pauli traveled from Hamburg, where he had just taken up his first faculty appointment.

Meeting Bohr for the first time was a turning point in the careers of Pauli and Heisenberg. The young Turks had much to learn from their wise elder and, as we have seen, a thing or two to teach him. Pauli, the older of the pair, was twenty-two; Heisenberg was only twenty. Sommerfeld had been a wonderful mentor, but now they encountered a different style in physics, a way of approaching problems that was unfamiliar to them, even when the material itself was not. Reflecting back on the meeting, Heisenberg said, "We had all of us learned Bohr's theory from Sommerfeld and knew what it was about. But it all sounded quite different from Bohr's own lips." It wasn't just the equations. As with a musical virtuoso, whose performance is more than just the playing of notes, the phrasing and the emphases mattered. One learned from Bohr more than simply what the equations said.

Again quoting Heisenberg, Bohr's

> insight into the structure of the theory was not the result of a mathematical analysis of the basic assumptions, but rather of an intense occupation with the actual phenomena, such that it was possible for him to sense the relationship intuitively, rather than derive them formally.
>
> Thus I understood: knowledge of nature was primarily obtained in this way, and only as the next step can one succeed in fixing one's knowledge in mathematical form and subjecting it to a complete rational analysis. Bohr was primarily a philosopher, but he understood that natural philosophy in our day and age carries weight only if its every detail can be subjected to the inexorable test of experiment.

Sommerfeld, himself a mathematician by training, believed in applying his formidable range of analytical techniques to formulating a problem prior to looking for the solution. Unlike Sommerfeld, Bohr struggled simultaneously toward defining a question and answering it. Doing so, he relied on a mixture of his own intuition and hints from experimental data. Unlike Sommerfeld's, which valued clarity above all, Bohr's lectures attempted to convey to his audience how hard he was searching for answers to questions and how much he needed others to join him in the quest.

The combination of Bohr's charisma and the beautiful Göttingen late spring was captivating, enveloping, never intimidating. At the end of Bohr's third lecture, Heisenberg stood up and asked some questions about a calculation that Bohr had been describing. The criticism he offered was very much to the point, and Bohr, somewhat taken aback, hesitated in answering. Right after the lecture he invited the twenty-year-old to go for a walk with him in the surrounding hills.

They walked and talked for three hours, discussing not only the details of calculations but also what it meant to understand a theory. As they walked, the conversation flowed over a wide variety of topics. Despite an age difference of almost twenty years, the two seemed to speak the same language. Bohr recognized Heisenberg's brilliance and daring, and the young German was both dazzled by the depths of Bohr's commitment to physics and flattered by Bohr's interest in him. Many years later Heisenberg reminisced about that day.

> This walk was to have profound repercussions on my scientific career, or perhaps it is more correct to say that my real scientific career only began that afternoon.

And in an interview, he recalled,

> It was my first conversation with Bohr, and I was at once impressed by the difference in his way of seeing quantum theory from Sommerfeld's way. For the first time I saw that one of the founders of quantum theory was deeply worried by its difficulties.

Bohr invited Heisenberg to come to Copenhagen, but circumstances delayed the visit for a year and a half. When Heisenberg finally arrived—

Easter 1924—Bohr was very busy. He was the head of a growing research establishment and the father of five young sons, but he made time for Heisenberg: "After a few days he came into my room and asked me to join him for a few days' walking tour through Sjælland. In the Institute itself, he said, there was little chance for lengthy talks and he wanted to get to know me better. And so the two of us set out with our rucksacks."

The walk took several days. They covered a hundred miles, walking north to Elsinore, the site of Hamlet's castle, and then circling home along the seashore. Conversations ranged widely, but in the end they were always about the mysteries of quantum theory. A bond had been forged. In September 1924, Heisenberg returned to Copenhagen for a seven-month stay, and in May 1926, for a full year, the crucial year during which he and Bohr hammered out the Copenhagen interpretation of quantum mechanics.

The Göttingen meeting also changed Pauli. He later said, "A new phase of my scientific life began when I met Niels Bohr personally for the first time." The relationship between him and Bohr never had the intense highs and lows that characterized Bohr and Heisenberg's. One cannot imagine Bohr and Pauli hiking together along the coast of Denmark, almost father and son, master and disciple, but one also cannot see them having discussions so heated that one of them has to leave the room. And neither Bohr nor Heisenberg would ever value the other's criticism as much as each did Pauli's.

Triumph and Crisis

A HIGH POINT of one of Bohr's 1922 Göttingen lectures was his discussion of element number 72 in the periodic table, seventy-two electrons orbiting around the nucleus. This turned out to be a key test of Bohr's predictive ability, of his *mystical magical wand.* In 1921, according to the by-then-accepted notion that elements are labeled by atomic number ranging from 1 to 92, three gaps in the progression had not yet been identified. Like all other elements, these three—43, 61, and 72—should each be a member of a family of elements with similar chemical behavior. Looking in places where other members of their family were found was the likeliest way to discover them.

In 1921, unbeknownst to Bohr at the time of the Göttingen meeting, two French scientists announced that they had found traces of element 72 in ytterbium samples, ytterbium being a member of the Mendeleev group known as the rare earth family. They promptly named the new element *celtium*, in honor of the Celts, the old inhabitants of France. The problem was that according to Bohr's scheme, 72 should not belong to that family.

Bohr first heard of the French result after returning from Göttingen. When Bohr wrote his German hosts to thank them for their hospitality, he added a postscript regarding his periodic table predictions:

> The only thing I know for certain as yet about my lectures in Göttingen is that several of the results I reported there are already wrong. A first point is the constitution of element 72 which, contrary to my expectations, has after all been shown by Urbain and Dauvillier to be a rare earth.

But what if the French had misidentified element 72? What if their experiment was wrong? According to Bohr, the place to look for element 72 was in zirconium, not ytterbium, samples. By 1922, Bohr's institute had a small experimental contingent, headed by Georg von Hevesy, his old friend from Manchester days. Encouraged by Bohr, Hevesy asked Dirk Coster, a young Dutch physicist who had pioneered a new X-ray detection technique, to join him in Copenhagen to verify the prediction. If Bohr was right, element 72 should not even be rare, much less a rare earth.

Hevesy and Coster quickly found it in zirconium samples, behaving exactly the way Bohr had predicted. It was a triumph for the institute. The element was now renamed *hafnium*, from Hafniae, the Latin name for Copenhagen. As icing on the cake, Bohr made the announcement of its discovery on December 11, 1922, in Stockholm, at the end of his speech accepting the 1922 Nobel Prize in Physics. The timing could not have been better. Coster was still triple-checking all the answers as Hevesy jumped on the Copenhagen-to-Stockholm train to attend the ceremony. (As a footnote, twenty-two years later Hevesy was himself the recipient of the Nobel Prize, in chemistry not physics, for his pioneering work with isotope tracers in biology.)

If Einstein had been with him in Stockholm, Bohr would have participated in a double ceremony, because the announcement and awarding of the 1921 physics prize was not made until 1922, and its winner was Albert Einstein, not for the theory of relativity but for his work in quantum theory. However, Einstein was away on a long trip to Japan when the announcement reached him. The German ambassador to Sweden accepted the prize in his place.

While in Japan, Einstein received a letter from Bohr describing what an honor it had been for him to be "considered at the awarding of the prizes at the same time as you." The Dane went on to say how grateful he was that Einstein had received the prize before him. Einstein, amused, wrote back a letter that begins affectionately, addressing his friend as "Dear or rather much more beloved Bohr" *(Lieber oder vielmehrgeliebter Bohr)* and includes the following:

> I can say without exaggeration that it pleased me as much as the Nobel Prize. I find especially charming your fear that you might have gotten the prize before me—that is truly "Bohrisch." Your new investigations of the atom have accompanied me on the trip, and they have made my fondness of your mind even greater.

Like Bohr, Einstein had also become an ambassador for science, espousing the view that it should ignore national boundaries. But his road was a more difficult one to follow than Bohr's because of his ambiguous position as both German and Swiss, the former by birth, the latter by citizenship. This was even reflected in the delicate dance surrounding the presentation of his Nobel Prize, accepted in his place by the German ambassador but given to him in Berlin by the Swiss ambassador.

Bohr's acceptance of the Berlin invitation in 1920 and his lectures at Göttingen had been intended as an affirmation of his view that science transcends national boundaries. At the 1922 Nobel Banquet, he proposed a toast "to the vigorous growth of the international work on the advancement of science, which is one of the bright points in human existence in these, in so many respects, sorrowful times."

They certainly were sorrowful times in Germany. Despite the Weimar government's efforts, the country's failing economy went from disastrous inflation to even worse hyperinflation. People rushed out to buy food as soon as they received their wages. Restaurants would not even print menus, because the price of a meal changed between the time it was ordered and when it was eaten. By September 1923 it took a million marks to buy what had formerly cost just one. Max Planck, leaving home to give a lecture, found himself having to sit up all night in a railroad station because he lacked funds for a hotel. By November 1923 that million had become a trillion. The hyperinflation could not be checked.

Hitler now made his move. On November 8, 1923, he burst into a meeting commemorating the end of World War I. He and his followers announced to the assemblage in a Munich beer hall that they would spearhead a march on Berlin to overthrow the *corrupt Weimar regime.* Their attempt failed quickly, Hitler was imprisoned, and there he turned to writing *Mein Kampf (My Struggle).* The Nazi endeavor was soon dismissed as one more right-wing fantasy.

Meanwhile atomic theory was undergoing its own crisis. After the Bohr model's spectacular success in explaining the details of hydrogen and ionized helium, it seemed logical to examine ordinary helium, number 2 in the periodic table, the atom with two electrons. Were the two electrons in the same orbit, or were they in different orbits inclined at an angle with respect to each other? Bohr and Kramers examined this problem, Born and Heisenberg reexamined it, but all their conclusions were unsatisfactory. Sommerfeld summarized the situation in 1923: "All attempts made hitherto to solve the problem of the neutral helium atom have proved to be unsuccessful." With the glow of the 1922 Göttingen lectures fading, an air of indecision developed. Though there had to be some ingredients, they were all missing, and nobody seemed to know how to proceed.

In January 1924 Bohr and Kramers made a proposal in collaboration with John Slater, the first of many Americans to come to Bohr's institute after receiving their PhD in America. Their thinking appeared in a twenty-page paper, really little more than a lengthy manifesto. It had no equations at all. Admittedly Bohr had a tendency to be wordier than other physicists, a reflection of his need for careful phrasing with all possible caveats, but no equations was extreme.

BKS, as the paper was soon called, attempted to overthrow many of physics' most sacred principles. Its first theme was that energy conservation and momentum conservation are not true in single subatomic collisions. The two principles only appeared to hold because experimental observations had always taken averages over many collisions, and what was true for the average was not true for the individual. The paper also claimed that strict causality, the apparent inexorable link between cause and effect, held only for the mean value taken over many measurements. BKS concluded by denying the existence of the photon, Einstein's proposed particle-like description of electromagnetic waves.

Einstein, who himself had earlier considered and then rejected many of the paper's suggestions, was adamantly opposed to every one of them. He didn't correspond directly with Bohr on the subject, but Ehrenfest heard from him that the idea of energy nonconservation was "an old acquaintance of mine whom, however, I do not regard as a respectable fellow" *(einen reellen Kerl)*, and that light quanta "can't be done without." Einstein wrote Born that if causality were abandoned, he would rather be "an employee in a gaming house than a physicist." The laws of physics, according to Einstein, had to have clear predictive value.

Pauli tried to act as an intermediary, communicating Einstein's objections to Bohr while explaining to him his own stance:

> Even if it were psychologically possible for me to form a scientific opinion on the grounds of some sort of belief in authority (which is not the case however, as you know), this would be logically impossible (at least in this case) since here the opinions of two authorities are very contradictory.

This disagreement was the beginning of the titanic disputes between the two physics giants. Bohr and Einstein, always so close in their respect and affection for each other, were from this point on seemingly always at loggerheads in their interpretation of subatomic physics.

Within a year experiments showed that photons did exist and behaved exactly as Einstein had predicted they would, thus resolving that issue. By early 1925, Bohr recanted. Trying to be graceful about BKS's failure, he wrote to one friend, "There is nothing else for us to do than give our revolutionary efforts as honorable a funeral as possible," and to another he declared that it "was very comforting that there is now no longer any reason to doubt the energy principle."

This first challenge to energy conservation had lasted less than a year. Bohr came back again to this same topic a few years later. This time his opponent was Pauli, not Einstein, and the subject of their disagreement would be played out in the "Copenhagen Faust."

The New Optimism

BOHR, KRAMERS, AND SLATER had been moved to make their suggestion in order to break out of the impasse quantum theory seemed to be caught in at the end of 1923, but in 1924 things began to look up in physics, as they also did in Germany. The country's economy took a rapid turn for the better when the government introduced the new Rentenmark, approximately equal to one prewar mark or one trillion of the paper marks then in circulation. Theoretically backed by a mortgage on German lands and goods, it was generally accepted, and the ravaging three-year inflation that had destroyed pensions and erased savings slowly came under control.

Fueled by a desire never to see a repeat of World War I's carnage, a general optimism about the resolution of political conflicts began to appear, at first tenuously, but quickly gathering strength. When the fourth Solvay conference was held in Brussels that year, the only German invited was Einstein. He declined, responding that the Allies' treatment of Germany was excessively harsh. But normalcy of relations soon set in. The general atmosphere of détente carried over into physics, and from then on Germans were once again invited to major international meetings. Fruitful exchanges between scientists of nations that had once been enemies now increased rapidly.

The year 1924 was a particularly good one for Pauli. He was happily settled in Hamburg, surrounded by a group of stimulating colleagues, many of them bachelors and bons vivants. With prosperity returning to Germany, restaurants were flourishing, and cabarets were packed. Hamburg was a big city, a port with clubs and a large red-light district; as mentioned earlier, it was in the part of town called, amusingly enough, Sankt Pauli. The young physicist began to enjoy himself.

That year also marks the beginning of the notion that Pauli should not be admitted to any physics laboratory lest a crucial piece of apparatus inexplicably break. A collection of stories, catalogued under the rubric of "The Pauli effect," began to circulate among physicists. Otto Stern, Hamburg's senior experimentalist, a good friend of Pauli's and his frequent luncheon companion, was the probable originator of the superstitious belief. Balding and jovial, the cigar-smoking Stern said that when he was in his laboratory, he would only speak to Pauli through a closed door for fear that the apparatus would break.

As the legend spread, examples of the superstition began to accumulate. A mishap in Copenhagen was attributed to his presence, and even a strange failure in a Göttingen lab was traced to Pauli's being on a train passing through town at the moment of the incident. Sometimes the Pauli effect could extend to car accidents, dishes falling off tables, and similar phenomena, but the common thread, which Pauli reveled in, was that he always remained unaffected as others suffered.

Despite often jokingly being asked not to enter a laboratory, Pauli studied experimental data closely. In 1923 he began a detailed analysis of radiation from atoms placed in a magnetic field. His aim was to understand how to assign quantum numbers to all the orbiting atomic electrons and thereby provide a complete and rigorous explanation of the periodic table of elements. Toward the end of 1924, he took a step in that direction, having realized that each electron needed four quantum numbers, one more than the original three that Sommerfeld had used to specify its motion. Those three were visualized as size, eccentricity, and inclination of orbit. The fourth one, having no immediate interpretation, took only two values, +½ and –½.

In order to explain the data, he next postulated the *exclusion principle*, sometimes simply known as the Pauli principle. It says no two electrons in the same atom can have an identical set of quantum numbers—at least one of the four has to be different.

In December 1924, Pauli sent Heisenberg a copy of a manuscript describing his conclusions. He received an answer almost immediately:

Dear Pauli!

Today, having read your new work, I am surely the one who is most pleased by it, not only because you have raised the swindle to unimagined, dizzying heights (by introducing single electrons with four degrees of freedom) breaking in the process all previous records that you accused me of holding, but quite generally am triumphant that you too (et tu, Brute!) have returned with bowed head to the land of formalism pedants, but don't be sad for you will be welcomed there with open arms. And if you have written something against earlier swindles, it's a natural misunderstanding; because swindle times swindle yields nothing correct so two swindles can never contradict each other. Therefore I congratulate you!!!!!!

Merry Christmas— Werner Heisenberg

As Pauli was sending his manuscript to Heisenberg, he received a warm note from Bohr in response to an earlier letter of his. It said, "My conscience is so bad that I am even ashamed to mention it. I had intended to write immediately to thank you for your long letter, which made me long so much for an opportunity to quarrel with you again." Touched by Bohr's evident affection, Pauli sent a copy of his four-quantum-number manuscript to him as well. A quick, characteristically supportive reply reached him a week after Heisenberg's. In it Bohr told Pauli how intrigued the Copenhageners were by the "many new beauties" he had brought them, continuing, "So you see, dear Pauli, that you have attained all you wanted, putting our thoughts in motion. I think we are now at a crucial turning point."

Within less than a year two young Leiden students of Ehrenfest's, Samuel Goudsmit and George Uhlenbeck, gave a possible interpretation of Pauli's fourth electron quantum number, though not in a form Pauli liked. Extending the planetary analogy, which Pauli no longer approved of, they pointed out that, just as the earth spins about its axis while it revolves around the sun, electrons might be able to spin while they orbit. Moreover quantum theory would constrain this spin to take on one of two possible values corresponding to two possible orientations: upward or downward.

Ehrenfest encouraged his students to write a short version of their idea in a form suitable for publication. A few weeks later they went to see Lorentz, the man who had brought Ehrenfest to Leiden, the dean of all European theoretical physicists. Now in retirement, he still lectured in Leiden from time to time and held discussions with students. Lorentz thought about their proposal for a while and then told them that though it was interesting, it contained serious flaws. Alarmed, the two students rushed back to Ehrenfest, saying they thought they should hold off submitting it for publication. He replied that he had already sent it, then added, "Well, you are both young, you can afford a stupidity." Shortly thereafter the validity of their conjecture became even more precarious. They received a letter from Heisenberg pointing out that their formulation of a spinning electron's effect did not have a crucial factor of 2 it needed to explain the data.

A few more weeks passed. In December 1925 a celebration in Leiden was set to honor Lorentz. Bohr and Einstein, both great admirers of the elder Dutchman, decided to attend the festivities. A story related to that meeting and to electron spin illustrates how small the world of theoretical physics was at the time. On the way from Copenhagen to Leiden, Bohr's train stopped in Hamburg. He found Pauli and Stern waiting for him, wanting to know what he thought of electron spin. He replied that the idea was interesting but apparently wrong. A day later, Ehrenfest and Einstein greeted him at the Leiden train station. They explained to him how, in the meantime, they had seen how the missing factor of 2 might be found, and how Lorentz's objections could be refuted by quantum theory. Bohr immediately switched to being an ardent advocate of electron spin.

A few days later, the celebration over, Bohr returned to Copenhagen with a stopover in Göttingen. Heisenberg was at the station, asking him what he thought about electron spin. He replied that it was a great advance, a triumph for quantum theory. Proceeding on, Bohr met Pauli at the Berlin train station, Pauli having made the trip from Hamburg expressly to see if Bohr had changed his mind during his Leiden visit. When he learned that he had, Pauli upbraided Bohr for sustaining the planetary analogy for the atom. But electron spin did stay, although suitably reinterpreted by quantum mechanics.

Meanwhile the Pauli principle was quickly generalized. We now know that two electrons in a system, be it an atom, a gas, a crystal, a magnet, or even a star, cannot have identical quantum numbers. This principle explains why all these different forms of matter are stable and why they retain their shape. The sun will end its life as a white dwarf star. Why don't electrons in a white dwarf star all collapse into the star's center? The Pauli principle forbids it. Why don't all the electrons in an atom jump down to the lowest energy state (think orbit if you wish, though Pauli would not approve)? The answer is, the Pauli principle forbids it.

The principle quickly became one of the two key ingredients of the new quantum theory. The other was a consistent way of assigning quantum numbers that did not depend on the visualization of orbits. One might well wonder, given the importance of his principle, why Pauli had to wait twenty years to receive the Nobel Prize. A large part of the reason may be that many physicists hoped for an explanation of why the principle held. Pauli himself provided a partial answer in 1940, but the Royal Swedish Academy of Sciences apparently still had reservations. These were finally overcome.

On January 13, 1945, Einstein sent the academy a telegram:

> NOMINATE WOLFGANG PAULI FOR PHYSICS PRIZE STOP HIS CONTRIBUTIONS TO MODERN QUANTUM THEORY CONSISTING IN SO-CALLED PAULI OR EXCLUSION PRINCIPLE BECAME FUNDAMENTAL PART OF MODERN QUANTUM PHYSICS BEING INDEPENDENT OF OTHER BASIC AXIOMS OF THAT THEORY STOP ALBERT EINSTEIN

That set the academy in motion. The Nobel Prize in Physics was awarded to Pauli on November 15, 1945, twelve years after Heisenberg and Dirac had received theirs. The citation said, "To Wolfgang Pauli, for the discovery of the exclusion principle, also named Pauli Principle."

Chapter 8

The Revolution Begins

CHORUS
Gird thee for the high endeavor;
Shun the crowd's ignoble ease!
Fails the spirit never,
Wise to think, and prompt to seize.

Goethe, *Faust, Part II,* act 1, 50–53

Helgoland

BY THE BEGINNING OF 1925, Heisenberg and Pauli had become convinced that the picture of electrons orbiting about the nucleus could not be the literal truth of what the atom's interior looked like. It was by now more than a decade since Bohr had first advanced this notion, and though the picture had many successes, it also had many failures. Admittedly it described very well the radiation from hydrogen atoms, but it was a complete fiasco in its depiction of helium. With the enthusiasm of youth, Pauli and Heisenberg felt a completely new picture was needed, one that did away with orbits. It might incorporate many features of Bohr's existing atomic theory, but the approach would be novel. In a late 1924 paper, Pauli even removed all mention of orbits, explaining to Bohr in a December letter that he had done so because "energy and momentum . . . of . . . states are something much more real than 'orbits' . . . We must not, however, put the atoms in the shackles of our prejudices (of which in my opinion the assumption of electron orbits . . . is an example)."

Bohr, not surprisingly, was finding it difficult to renounce the representation he had pioneered, but Pauli brooked no compromise. When at Christmas 1924, Bohr expressed his enthusiasm for Pauli's recent pronouncement that no two electrons can have the same quantum numbers, Pauli's reply was characteristically severe: "Weak men, who need the crutch of defined orbits and mechanical models, can think of my rule as saying that electrons with the same quantum numbers would have the same orbits and therefore collide with one another." But what was the alternative for strong men? Who would lead them away from the safety of orbits, and where would the trip end? Sadly, it would not be Pauli.

Many years later he reflected wistfully:

> When I was young, I thought I was the best formalist of my time. I thought I was a revolutionary. When the big problems would come, I would solve them and write about them. The big problems came and passed by; others solved them and wrote about them. I was a classicist and not a revolutionary.

As always, Pauli was as harsh with himself as he was with others. Though Pauli is undervaluing here his own contribution, it is indisputable that Heisenberg, not he, was the daring innovator. Pauli may have had greater insight and critical ability, but he was not the intellectual risk taker that Heisenberg was.

In February 1924, two months before Heisenberg arrived for his long stay in Copenhagen, Pauli wrote Bohr a letter assessing his friend's abilities: "He does not pay attention to clear elaboration of the fundamental assumptions and their relation to existing theories. However . . . I believe that some time in the future he will greatly further our science." These words were prophetic. Heisenberg's ability and willingness to depart from "existing theories" were what led him to quantum mechanics.

Like Pauli, Heisenberg maintained that visualizing electrons as moving in orbits could not be part of the theory. One can *trace* the whole planetary orbit. By contrast, electrons might behave *as if* they move in orbits, but one cannot trace their paths. In a deeper sense Heisenberg's basic idea was that the theory should only involve quantities that are experimentally measurable—in the language of physics, they are known as *observables*—and electron orbits were not observable.

Physicists believe the quantum mechanics revolution proper began in mid-June 1925 in a very curious place, Helgoland, a tiny grassless island in the North Sea. Here's how it happened. Heisenberg had retreated to Helgoland from Göttingen, where he had been working since his return from Copenhagen. He was seeking relief from a hay fever attack, an affliction that had plagued him since youth. Fleeing the mainland for what he thought would be a short stay, he only brought along a few items of clothes, a pair of hiking boots to climb on the seaside rocks, a copy of Goethe's *West-Östlicher Divan (Poems of the West and the East)*, and some calculations he was having difficulties with. He hoped to work on them if he felt well enough.

Not having made progress in doing away with the orbit picture, Heisenberg decided to try looking at a simpler problem, thinking it might show him the way to proceed, a sort of test run before the big race. The problem he settled on was akin to the motion of electrons oscillating back and forth along a line rather than on the more realistic and also more complicated elliptical paths they were thought to trace out. If he could find a means of describing the radiation in his model without explicit knowledge of the electron trajectory, he hoped it might suggest to him how to do away altogether with orbits.

The key step was to discover a rule for multiplying together the amplitudes of the electrons' oscillatory motion. Heisenberg knew he was missing a crucial restriction that needed to be imposed on this multiplication. One night on Helgoland he found it, though he didn't realize until later the full significance of what it was that he had discovered.

> There was a moment in Helgoland in which the inspiration came to me . . . It was rather late at night. I laboriously did the calculations

> and they checked. I then went out to lie on a rock looking out at the sea, saw the Sun rise and was happy.

Cured of his hay fever, Heisenberg returned to Göttingen, stopping first in Hamburg to tell Pauli about the conclusions he had reached on the island. Wasting no time, he immediately started preparing a manuscript for publication. On the twenty-ninth of June, 1925, he wrote Pauli that he was "convinced in his heart that this new quantum mechanics is truly right," and on the ninth of July Heisenberg sent him the finished manuscript. It begins as follows and includes a quotation cited earlier:

> It is well known that the formal rules which are used in quantum theory for calculating observable quantities such as the energy of the hydrogen atom may be seriously criticized on the grounds that they contain, as basic element, quantities that are apparently unobservable in principle . . . In this situation, it seems sensible to discard all hope of observing hitherto unobservable quantities, such as the position and period of the electron, and to concede that the partial agreement of the quantum rules with experience is more or less fortuitous.

In plain language, physics needed a new theory. Was his proposal the first step toward it?

Another Sleepless Night

HEISENBERG PREPARED a second copy of the manuscript. Still not fully aware of the importance of what he had discovered, he gave this second copy to Max Born, Göttingen's senior theorist, asking him to evaluate the paper's contents and, if Born thought it was worth publishing, to please forward the manuscript to the *Zeitschrift für Physik*, the leading German physics journal. Heisenberg then set off to give a previously scheduled lecture in Cambridge.

Believing there was no urgency, Born put the manuscript away for a few days while he tended to other matters. When he finally read Heisenberg's paper, it was Born's turn for a sleepless night. He found the curious multiplication rule Heisenberg had discovered for the oscillation amplitudes strangely familiar. As Born remembers,

> I began to ponder about his symbolic multiplication, and was soon so involved in it that I thought the whole day and could hardly sleep at night. For I felt there was something fundamental behind it . . . And one morning . . . I suddenly saw light. Heisenberg's symbolic multiplication was nothing but the matrix calculation well known to me since my student days.

Born realized that Heisenberg's physics was original, but the mathematical techniques he had used had already been explored. In effect Heisenberg had rediscovered them. This meant the theorists didn't have to invent a whole new branch of mathematics to advance Heisenberg's ideas. Mathematicians had already written books that laid out the necessary procedures. The physicists only needed to learn how to use the new tools, not also how to build them. A somewhat similar situation had occurred a decade earlier when Einstein had formulated his general theory of relativity, relating the curvature of space to the presence of energy. The theory of differential geometry, developed over decades by mathematicians, had been the tool in that case, helping the field to advance much more rapidly than would otherwise have been possible

Born was very excited about his discovery and eager to share his enthusiasm with others, in particular with Pauli, who had been his assistant in Göttingen before Heisenberg. He also knew that Pauli was the one person other than of course Heisenberg and himself who was com pletely familiar with this new work. On July 19 the two of them met by arrangement on a train they were taking to a meeting in Hannover. Born recollects what happened next:

> I joined him in his compartment, and absorbed by my new discovery, I at once told him about the matrices and my difficulties in finding the values of those non-diagonal elements. I asked him whether he would like to collaborate with me in this problem. But instead of the expected interest, I got a cold and sarcastic refusal. "Yes, I know you are fond of tedious and complicated formalism. You are only going to spoil Heisenberg's physical ideas by your futile mathematics."

Pauli was wary of mathematical formalism even though it came relatively easily to him. He wanted neither Bohr's "crutches for weak men" nor Born's "futile mathematics." Writing to a friend, Pauli said Heisenberg's work had restored his zest for life. However, still suspicious of Born, he added, "One has to ensure the work isn't all going to be covered up by a deluge of Göttingen erudition." Believing Heisenberg's bold proposal had freed quantum physics from "the shackles of prejudices," Pauli did not want this newfound freedom to be compromised by intrusions from elders. Though only twenty-five, Pauli had worked with Sommerfeld, Born, and Bohr. He had learned from them, appreciated them, but with a new era dawning, he was confident that the young would lead the way.

It turned out that Pauli was wrong about Born's "futile mathematics." Together with twenty-two-year-old Pascual Jordan and later with Heisenberg himself, Born found the essence of the strange new multiplication rule Heisenberg had discovered. Meanwhile twenty-three-year-old Paul Dirac independently discerned the same relation.

In learning arithmetic, we quickly come to appreciate that 3 times 4 is the same as 4 times 3. The commutative law of arithmetic says that *AB* equals *BA*, where *A* and *B* are any two numbers. Similarly, in classical mechanics as formulated by Newton, measurements of a particle's position and velocity give the same result regardless of the order in which the measurements are taken. Not so in the new quantum mechanics.

With *A* as the particle's position and *B* as its momentum (momentum, though it has a more sophisticated definition, can be thought of as mass times velocity), the physicists found that *AB* minus *BA* was proportional to Planck's constant. Why didn't the product of position and momentum commute? What difference did it make if one looked at velocity first and position second or position first and velocity second?

Could this strange multiplication rule explain the results of Bohr's model for the hydrogen atom without resorting to the picture of orbits? This was the great challenge that had pushed Heisenberg and Pauli forward in the first place. If it could, they would know they were well on their way to a radically new formulation of quantum theory. The revolution would have begun!

Pauli swung into action. In a tour de force he succeeded in calculating the hydrogen energy spectrum by using Heisenberg's new formalism. The result agreed, as hoped, with Bohr's 1913 result, the so-called Balmer formula. Pauli, however, in the paper he wrote describing this work, underlines the new technique's superiority by stating on page 1, "Heisenberg's form of quantum theory completely avoids a mechanical-kinematic visualization of the motion of electrons." Pauli goes on in the same paper to treat the effect of small perturbations due to external electric and magnetic fields, reaching successes that had proved elusive in the old quantum theory.

Enthusiastic about his newly obtained results, Pauli quickly informed his friends by letter of his conclusions. On November 3, 1925, almost by return mail, Heisenberg wrote back, "I don't need to write you how pleased I am by your new theory of hydrogen and how admiring I am that you could bring it out so quickly." Pauli had succeeded in doing all the calculations in just a few weeks. Ten days after that, Bohr echoed Heisenberg's sentiments: "To my great joy, I heard from Kramers that you succeeded in obtaining the Balmer formula. I am really interested in hearing about this and hope, as you promised Kramers, that you'll write to me about it very soon."

On January 17, 1926, Pauli submitted for publication his paper, entitled "On the Hydrogen Spectrum from the Point of View of the New Quantum Mechanics." Its appearance convinced physicists that

Heisenberg's was a new and important approach to the problems in understanding the atom. Quantum mechanics was now fully launched. Its birth had many midwives; Pauli, Dirac, Born, and Jordan each made major contributions, but Heisenberg had been, in Pauli's words, the *revolutionary*.

And then something else amazing happened, almost simultaneously. A thirty-eight-year-old Austrian named Erwin Schrödinger found a totally different way of solving the same problems. A few months earlier there had been no way in sight of avoiding the impasse blocking progress in atomic physics, and now there were two. Not realizing how close they were to one another, physicists gave them separate names. Heisenberg's was called matrix mechanics after the unfamiliar mathematical tools he was using, while Schrödinger's was wave mechanics, because his electrons seemed to be guided by waves.

Simultaneous discoveries are not uncommon in science, but important simultaneous discoveries that approach a problem from different directions are very rare. It is as if two parties, strangers to each other, reached the summit of a previously unclimbed Himalayan peak at the same time, one approaching from the east and the other from the west.

The physics community, in a state of both shock and excitement, was soon informed that matrix mechanics and wave mechanics were the same theory, but the scientists required several years to fully understand its implications. The starting points of the two mathematical approaches were different and seemingly unrelated, but the endpoints were the same.

Waves or Particles

SCHRÖDINGER HAD SET OFF on his quest after learning of a young Frenchman's idea, one so simple that many other theoretical physicists must have wondered why they hadn't thought of it. It wasn't technically innovative like the advances of Heisenberg, Pauli, or Dirac. It was simply lying there, waiting to be gathered in, the kind of discovery that gives ordinary theoretical physicists hope. You may not be a towering genius, but sometimes you can be lucky. It usually occurs only once in a lifetime, but once is enough. If it happens more than once, it probably was not luck in the first place.

Whatever the mixture, skill and luck came together in 1923 for this unlikely candidate, Louis-Victor, Prince de Broglie. At thirty-one, he was living in the family's Paris mansion with his older brother, Maurice, Duc de Broglie, their parents having passed away years earlier. With a carefully trimmed mustache and languid appearance, dressed formally with a winged collar, Louis de Broglie looks in pictures of the time every inch, or rather every centimeter, the French aristocrat that he was. He originally intended to join the diplomatic service, but World War I intervened and his interests changed. At twenty-eight he enrolled as a graduate student in physics, thinking he might engage in the same profession as his brother, who was a reasonably well-known physicist with an independent laboratory housed in the family mansion.

Four years later Louis de Broglie was ready to submit his thesis to the Sorbonne. The faculty did not quite know what to make of it, but they were not inclined to fail someone so socially prominent. They passed him, praising him for his commendable efforts, though some mockingly called his work a Comédie-Française. However, one of the examiners, Paul Langevin, thought there might just be some profound truth in the thesis. Knowing Einstein from conferences, he sent him a copy. After studying it, Einstein wrote Lorentz, "I believe it is a first feeble ray of light on this worst of our physics enigmas."

In 1929, de Broglie was awarded the Nobel Prize for having provided that "first feeble ray of light," and it soon went as well to the two experimentalists who had verified his conjecture. The Swedish academy's tribute to him in 1929 concludes,

> When quite young you threw yourself into the controversy raging round the most profound problem in physics. You had the boldness to assert, without the support of any known fact, that matter had not only

> a corpuscular nature, but also a wave nature. Experiment came later and established the correctness of your view. You have covered in fresh glory a name already crowned for centuries with honor.

It was all quite simple. De Broglie knew that in 1905 Einstein had maintained that electromagnetic radiation travels as packets of energy, quanta, with each quantum having a particle-like nature. De Broglie now simply turned this argument around. If light is made of photons and therefore waves are particles, then particles should be waves. More to the point, electrons ought to display wavelike behavior. In de Broglie's own words,

> After long reflections in solitude and meditation, I suddenly had the idea during the year 1923, that the discovery made by Einstein in 1905 should be generalized by extending it to all material particles and notably to electrons.

What remained was obtaining a formula for the wavelength of a moving electron. Einstein had used his theory of relativity to derive one for the photon's momentum in terms of the radiation wavelength. De Broglie now simply reversed it, finding the electron wavelength in terms of its momentum.

He then tried to see if his formula could predict Bohr's quantization of atomic electron orbits by thinking of orbits as cords, vibrating while waves move back and forth on them. Pythagoras had shown 2,500 years earlier that stretched cords could only sustain notes whose wavelengths have a certain simple relation to the cord's length. But if wavelengths are inversely proportional to momenta, as de Broglie maintained, only certain momenta are possible on the cord/orbit. This led to the totally new derivation of Bohr's quantization rule.

De Broglie's idea also had quickly realized practical consequences. A microscope's essence is the focusing of a light beam shining on an object. Since electrons are also waves, an electron microscope could now be built by focusing electron beams. A microscope's ultimate limitation is the beam's wavelength, because objects smaller than it cannot be visually separated from one another. The best optical microscopes magnify by a factor of one thousand, but since an electron's wavelength can be made so much smaller than that of visible light, electron microscopes today enlarge by as much as 100,000. The insides of cells are too minute to be seen by the former, but the latter can examine them in great detail. In this way an esoteric application of quantum theory was quickly seen as capable of changing how chemistry and biology are studied.

De Broglie's great discovery may simply have been a case of a once-in-a-lifetime flash of intuition with a good deal of luck. There are no such doubts about Schrödinger, though his initial dramatic success was almost as surprising as de Broglie's. The only child of a solidly bourgeois Viennese family, Schrödinger had been a star pupil in the *Gymnasium* and later at university, but his career was interrupted by a year of compulsory military service in 1910 and then by four years as a soldier in World War I. At the end of the war he was back at the University of Vienna, a thirty-year-old trying to get his career going in what had become a second-rate physics research institute, one whose best students, like Pauli, were going elsewhere. Schrödinger's work was respectable enough to enable him to climb the academic ladder with positions in succession at the universities in Jena, Stuttgart, and Breslau; in 1921 he was appointed to a professorship of theoretical physics at the University of Zurich, where for the first time he had colleagues of the highest caliber.

However, Schrödinger was a solitary worker, not wanting students or collaborators. By the fall of 1925 he had still not produced any world-class research, and it seemed unlikely that he would start at the age of thirty-eight. Once thinking he was destined for greatness, Schrödinger saw his dream rapidly fading. He knew as well as anybody that Einstein and Bohr had become major figures in their midtwenties and that Pauli and Heisenberg were following in their footsteps.

In addition to career problems, his marriage seemed to be on the rocks. Never a faithful husband, he was carrying on a series of liaisons, and his wife, Anny, was having an affair of her own. Their friends thought the marriage was about to end.

At Christmas 1925, Schrödinger went off on a two-week vacation to Arosa, an Alpine resort in Switzerland where he had previously convalesced from a touch of tuberculosis. His companion was an unknown woman, perhaps an old friend from Vienna—her identity remains a mystery. Was it her inspiration that led to Schrödinger's remarkable breakthrough to greatness? Schrödinger's Zurich colleague Hermann Weyl once remarked, "Schrödinger did his great work during a late erotic outburst in his life."

All we know for sure is that Schrödinger returned from Arosa with the equation that forms the basis for wave mechanics, the equation that students of chemistry and physics still spend endless hours examining. He clearly had been thinking about quantum theory for a while, and we know he had studied de Broglie's papers, but the ultimate denouement arrived quickly, just as it had for Heisenberg, though the surroundings in which it came form a striking contrast to Helgoland's barren shores on which Heisenberg had pondered in solitary splendor.

In the six months that followed, Schrödinger wrote four papers, successively applying solutions of his equation to an astounding number of problems: the hydrogen atom, the diatomic molecule, the harmonic oscillator, perturbations due to external electric and magnetic fields, and absorption and emission of radiation. It quickly became clear to theoretical physicists that understanding the nature of atomic and molecular binding lay in solving this equation in new settings. Furthermore, in order to have mathematically consistent solutions, the equation demanded that certain parameters equal one of a constrained set of values, known as *eigenvalues*, quickly shown to be the same as Bohr's quantum numbers. It was all fitting together.

Schrödinger's four papers, "Quantization as an Eigenvalue Problem I, II, III and IV," read like a modern standard first-year graduate quantum mechanics course. During this six-month period Schrödinger also wrote a fifth paper, "On the Relation of the Heisenberg-Born-Jordan Quantum Mechanics to Mine," displaying the equivalence of the two formalisms. But the contrast between the two could not have been greater. Matrix mechanics seemed ad hoc, obscure, and hard to grasp, employing unfamiliar tools. By contrast the essence of Schrödinger's work was a well-defined mathematical formalism for de Broglie's particle-waves. Schrödinger used techniques that were familiar to physicists, ones they had studied in advanced calculus courses.

A good equation for a physicist is like a hammer for a carpenter. Schrödinger's equation, more akin to a marvelous new toolbox, gave the means to solve whole varieties of problems. Even today, much of physics is summarized in Newton's relation between force and acceleration, $F = ma$; Maxwell's four equations for electricity and magnetism; and Schrödinger's equation for quantum mechanics. Go into any physics or chemistry classroom in the world and you will find students still marching through this progression.

Schrödinger's work in the first half of 1926 has been considered an extraordinary example of both creativity and productivity, rivaled in this century perhaps only by Einstein's dramatic contributions of 1905. He solved in a few months many of the key problems in quantum theory, displaying mathematical virtuosity and physical insight as he progressed. Even though nothing he did after this reached the same peak, there would never be a question of the depth of his thinking. Luck entered into his success, as it always must, but there was much, much more to his achievement.

Heisenberg versus Schrödinger

Even before the last in his string of five seminal papers appeared, Schrödinger was hailed as the new messiah of physics. Planck wrote him in April 1926 that he read the first papers "the way an inquisitive child listens in suspense to the solution of a puzzle that he has been bothered about for a long time." Einstein sent Schrödinger a letter saying, "The idea of your article shows a real genius," and a few days later a second one, "I am convinced that you have made a decisive advance with your formulation of the quantum condition, just as I am equally convinced that the Heisenberg-Born route is off the track." Ehrenfest let Schrödinger know that "every day for the past two weeks our little group has been standing for hours at a time in front of the blackboard in order to train itself in all the splendid ramifications" of the equation.

It seemed that all the physicists had to do was study the marvelous equation and quantum theory's problems would disappear. When Planck retired from his professorship in Berlin at the end of 1926, Schrödinger was chosen as his replacement. Theoretical physics had never seen such an immediate, clamorous success.

But not everybody was happy. In late April 1926, at the height of the excitement, Heisenberg came to give a scheduled lecture on matrix mechanics to the Berlin physics faculty and their students. The knowledgeable elders found Heisenberg's formalism baffling. Three months later, in July, the distinguished Berliners were treated to Schrödinger's view of quantum mechanics, much more to their liking. He was one of their own, not a twenty-four-year-old *boy* challenging and confusing them.

Journeying triumphantly home from Berlin to Zurich, Schrödinger stopped in Munich to present his findings. Coincidentally, Heisenberg was visiting his parents and came to the lecture. As it ended, he rose to object to Schrödinger's interpretation of quantum mechanics, emphasizing that it involved unobservable quantities. The audience's sympathies were, however, not with him, even in his hometown. One of the Munich senior professors interrupted him firmly: "Young man, Professor Schrödinger will certainly take care of all these questions in due time. You must understand we are now finished with all that nonsense about quantum jumps."

Nor was Schrödinger shy about the contrast between the two new approaches. In the last of his five 1926 papers, the one in which he directly compares wave to matrix mechanics, he has a footnote explaining how he arrived at his conclusions:

> My theory was inspired by L. de Broglie and by brief but infinitely far-seeking remarks of A. Einstein. I was absolutely unaware of any generic relationship with Heisenberg. I naturally knew about his theory, but because of the very difficult-appearing methods of transcendental algebra and because of the lack of visualizability, I felt deterred by it, if not to say repelled.

Repelled is a very strong term for the usually emotion-neutral language of physics papers. I have never seen it used in any similar published work. But Schrödinger meant what he was saying. Perhaps unwittingly, he was setting up a confrontation, one that Heisenberg would not back away from. A battle was taking shape.

Schrödinger may have been repelled by Heisenberg's matrix mechanics, but Heisenberg's views were no less strong. He was convinced that Schrödinger's picture was incorrect: an electron could not have its motion determined by a guiding wave, the object that Schrödinger had given the name *wave function* or *psi function*.

The question was not whether matrix mechanics or wave mechanics was valid. They were equivalent, as Schrödinger and Pauli had independently shown. The meaning of Schrödinger's version of quantum mechanics is what puzzled Heisenberg. Schrödinger favored the view that one should do away with the idea of atomic electrons jumping from one orbit to another. The solution to his equation, the wave function, was for him the key, the rudder steering the electron's motion. Heisenberg, on the other hand, thought the jump was the only observable. According to his point of view, Schrödinger's wave function was not a physical observable and therefore not measurable.

At the end of June 1926, Max Born submitted for publication a paper entitled "Quantum Mechanics of Collision Phenomena." He thought he had broken the Gordian knot by finding an interpretation that adjudicated some credit to both contenders. Schrödinger's wave function was indeed unobservable, as Heisenberg maintained, but the magnitude of its square (technically absolute value squared) was observable—it could be related to the probability of finding an electron. However, and this was the key, no absolute determination of an electron's position was possible, only relative probabilities of detecting it in one position or another.

Although Born received the Nobel Prize thirty years later for the paper's contents, Schrödinger and Einstein were no more pleased by this formulation than they had been by Heisenberg's. It abandoned the notion of causality, the cherished link for them between cause and effect. One of Einstein's most famous quotes is contained in a December 1926 letter on the subject to Born. He simply would not accept the notion that a probability measurement of an electron's position was the best one could do:

> Quantum mechanics is very impressive. But an inner voice tells me that it is not yet the real thing. The theory produces a good deal but hardly brings us closer to the secrets of the Old One. I am at all events convinced that He does not play dice.

Despite Born's help, the tide seemed to be turning against Heisenberg. Fortunately for him, Bohr was coming to the same conclusions as he. Schrödinger may have been the messiah, but he would now have to do battle with "the Lord."

In May 1926, at the height of the crisis, Heisenberg moved from Göttingen to Copenhagen to take up a prearranged position as Bohr's assistant. It had become vacant after Kramers, who had held it for many years, returned to the Netherlands to assume the professorship of theoretical physics at the University of Utrecht. Heisenberg had hesitated before accepting Bohr's offer because he had been proposed for an associate professorship in Leipzig, and German rules penalized candidates for declining a valid offer. His father, remembering his own struggles, urged his son to accept the Leipzig position and the security that went with it, but young Heisenberg was more interested in advancing quantum mechanics than in looking for security. Also, Born and others assured him that he need not worry about climbing the academic ladder—a full professorship in a German university would be his soon enough.

Heisenberg went to Copenhagen and moved into the newly renovated small flat on the third floor of Bohr's institute. The increasing number of physicists coming to Copenhagen and the growing size of the Bohr family had made fitting everybody into the single Blegdamsvej building impossible. As usual, Bohr had taken action. In 1924 ground was broken for an adjacent second building, where the family would live. This meant more room for theoretical physicists and the availability in the old building of a guest suite, now occupied by Heisenberg. This friendly and intimate atmosphere provided Heisenberg with the support he needed to concentrate his thoughts and begin formulating his opposition to Schrödinger's point of view. Now a virtual member of the Bohr family, he was free to come and go into their house, to play their piano and interrupt whenever he wanted.

Conversely, Bohr often walked over to Heisenberg's rooms:

> After eight or nine o'clock in the evening, Bohr, all of a sudden, would come up to my room and say, "Heisenberg, what do *you* think about this problem?" And then we would start talking and talking and quite frequently we went on until twelve or one o'clock at night.

Bohr and Heisenberg had never been closer, and Pauli, in nearby Hamburg, kept in frequent communication with both of them. The three now joined forces as the key opponents to the growing acceptance of Schrödinger's views.

In October 1926 Schrödinger arrived in Copenhagen to present the arguments he had used to sway the Berlin and Munich audiences. This time the reception was less sympathetic. As described earlier, the argument between Bohr and Schrödinger, two battling titans, went on from morning till night, not stopping even when Schrödinger became sick and retired to his bed. Neither of them was able to convince the other to change his viewpoint. In the end they came away with no agreement on quantum mechanics, but with renewed respect for each other, even though they had often seemed to be talking past, rather than to, one another. When he finally returned home, Schrödinger wrote to a friend saying what a remarkable man Bohr was, but how impossible he found it to argue with him: "You no longer know whether you take the position he is attacking, or whether you must really attack the position he is defending."

Bohr, Heisenberg, and Pauli felt that quantum mechanics was less like classical mechanics than Schrödinger was willing to concede, and Bohr would simply not settle for anything less than complete clarity. As Heisenberg remembered many years later,

> Most other physicists are inclined to stop somewhere and say, "All right, we have it." But Bohr would never do it that way. Bohr would follow the thing out to the very end, just to the point where he was at the wall.

Schrödinger's equation had provided "mathematical clarity and simplicity" to quantum physics, but the question being asked in Copenhagen was, What did it really mean? Were there, perhaps, intrinsic limits to what one could describe? Might they be contained in that essential step that Born and Jordan and, independently, Dirac had emphasized, the notion that momentum p and position q did not commute, that $pq - qp$ is not zero, that it is instead proportional to Planck's constant? If p is mass times velocity, this said something about the simultaneous measurement of a particle's position and velocity, but what was it that it said?

In late October 1926, Pauli wrote Heisenberg a long letter, asking him why this was so.

> One may view the world with a p-eye and one may view it with a q-eye, but if one opens both eyes at the same time, one goes crazy . . .

I am really looking forward to your answer. Now, for once, you can criticize and make rude remarks.

Pauli, the eternal critic, egged Heisenberg on, saying that one cannot "inquire simultaneously about p and q." He received an answer quickly. Heisenberg told him how very enthusiastic he was about Pauli's letter and that it had made the rounds with Bohr, Dirac (then visiting Copenhagen), and others fighting over how to reply to his comments. Unfortunately that is the last letter we have from Pauli to Heisenberg until January 1933. The 1927–32 letters were lost during World War II, and Pauli had not kept copies.

There was a general agreement in Copenhagen that Schrödinger's interpretation of quantum mechanics was inadequate, the group boiling the argument down to a few key issues. How does one measure where an electron is and how it moves? Beyond that, what does measurement even mean in the atomic environment? Is an electron a particle, a wave, or both? What is the essential difference between the classical world and the quantum world? Up until now, there had been no real tension between Bohr and Heisenberg, but the two of them began to forcibly disagree with each other as the struggle for a final understanding of quantum mechanics reached its climax. How to proceed? What notions should guide them? Having left behind their differences with Schrödinger, the battle now increasingly pitted Bohr against Heisenberg. So close only a short time earlier, they no longer could find common ground.

Heisenberg, adamantly opposed to Schrödinger's point of view that electrons move according to a "guiding wave," felt that the mathematical formalism should be their guide, with questions of how to interpret experiments coming later. Bohr maintained that although Schrödinger's "guiding wave" was not completely correct, it had to be part of the synthesis. First they must obtain an overall understanding of the physical situation. Mathematics would come later.

Several theoretical physicists played important roles in the intense discussions that went on daily through the last months of 1926 and into 1927. Even though many others participated, in the end the principals were always Bohr and Heisenberg, "the Lord" and the new messiah. They argued passionately day after day until they were both exhausted. I'm not sure of this, but I imagine they must have sensed they were reaching the decisive point in their ongoing discussions, and neither was willing to give way.

From time to time Bohr's close friend Oskar Klein tried to intercede. Kramers and Klein had been Bohr's two co-workers at a time when there wasn't even an institute, and Klein had now returned to Copenhagen after some years in the United States. Kramers warned him, "Do not enter this conflict, we are much too kind and gentle to participate in that kind of struggle. Both Bohr and Heisenberg are tough, hard-nosed, uncompromising and indefatigable. We would be crushed in that juggernaut." There was only one person who could safely come between them: Wolfgang Pauli.

Uncertainty and Complementarity

BY MID-FEBRUARY Bohr and Heisenberg were at loggerheads, worn out, fighting with each other. Heisenberg later described their fatigue:

> Both of us became utterly exhausted and rather tense. Hence Bohr decided in February 1927 to go skiing in Norway, and I was quite glad to be left behind in Copenhagen where I could think undisturbed about those hopelessly complicated problems.

It was supposedly a vacation, but Bohr and Heisenberg both knew they needed to separate for a while. Within a few days of Bohr's departure, Heisenberg found what he had been looking for. It came to be known as the *uncertainty principle*, one of the greatest and still most puzzling advances of twentieth-century science. Formulating a first version of his conclusions in a fourteen-page missive to Pauli, he ends the letter, "I know very well this is still unclear in many points, but I have to write you in order to make it somewhat clearer. Now I await your merciless criticism." Heisenberg wanted to make sure his own thinking was correct, but also, knowing the discussions with Bohr were not going to be easy, he wanted help from the one person he was sure Bohr would listen to. Pauli studied the letter and approved.

Heisenberg had a manuscript ready when Bohr returned from Norway in the middle of March. Bohr quickly realized that Heisenberg's result was very important, but in his view the paper was not more than an interesting first draft. He thought Heisenberg still had not dealt adequately with the particle-wave dichotomy of submicroscopic particles, such as electrons, and should not publish anything until he had a better understanding of what his new result implied.

Heisenberg, on the other hand, wished to publish immediately. Realizing that Schrödinger's approach to quantum mechanics was becoming dominant, Heisenberg wanted physicists to know at once that his new result undercut Schrödinger's premise that an electron's position and velocity could be simultaneously specified.

The arguments began again. Heisenberg recalled, "Bohr tried to explain that it was not right and I shouldn't publish the paper. I remember that it ended by my breaking out in tears because I just couldn't stand the pressure from Bohr." But he did resist, and left the paper unchanged, except for a note that mentioned forthcoming work by Bohr. On March 22, 1927, he submitted it for publication.

No matter how stressful the situation had become, Bohr admired Heisenberg's mental toughness. Bohr's biographer Abraham Pais remembered a conversation in which "Bohr and I talked about an event when a senior theoretical physicist had talked a younger colleague out of publishing a result that turned out to be correct and important. When I remarked that this was a sad story, Bohr literally rose and said: 'No, the young man was a fool.' He explained that one should simply never be talked out of anything that one is convinced of." Perhaps Bohr thought back to his own reaction when Rutherford had written him, suggesting he make a few changes in his 1913 paper about the hydrogen atom. He had gone immediately from Copenhagen to Manchester and explained to Rutherford that nothing in the paper could be changed.

The conflict between Bohr and Heisenberg was finally resolved in early June 1927 by the eagerly awaited arrival in Copenhagen of Pauli. He made the two antagonists realize that they only disagreed on the order in which they should proceed, not on anything substantive. Resting on Heisenberg's uncertainty principle and Bohr's complementarity principle, this is the beginning of what came to be known as the Copenhagen interpretation of quantum mechanics.

Uncertainty principle may be an unfortunate name since it suggests that, with some decisiveness, one could do better. Sometimes, rarely, "indeterminacy" is used in technical discussions, but the awkward-sounding "unknowability" or "impossibility" is closer to the principle's true meaning. What Heisenberg showed was that if one accepts quantum mechanics' validity, certain types of measurements are impossible. Classically, one can simultaneously specify the position of a particle and its momentum; quantum mechanics says one cannot. The best one can do is a measurement for which the unknown part of the position measurement times the unknown part of the momentum measurement equals Planck's constant.

Heisenberg also showed how causality is subtly invalidated. A straightforward violation would deny that the present determines the future. His principle, on the other hand, asserts that we cannot predict the future because we cannot know the present with arbitrary precision. In his own words,

> In the sharp formulation of the causality law: "If we know the present, then we can predict the future," it is not the consequence, but the premise that is false. As a matter of principle we cannot know all determining elements of the present.

1. Portrait of Johann Wolfgang von Goethe in the Roman Campagna, by his friend Johann Tischbein, 1787.

2. Lise Meitner standing in a Berlin garden, c. 1910.

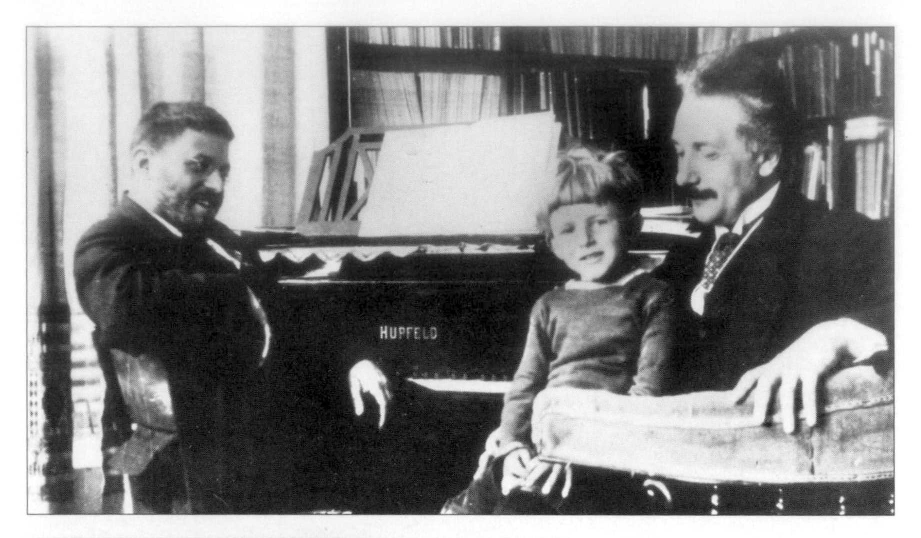

3. Albert Einstein with Paul Ehrenfest and his son Paul Jr. in Ehrenfest's Leiden home, 1919.

4. Einstein and Niels Bohr walking on a Berlin street, 1920.

5. Bonzenfreie Kolloquium, Berlin, 1920. Meitner is at the center of the picture with Bohr to her right and Otto Hahn directly behind her.

6. Niels and Margrethe Bohr in the early 1920s.

7. Enrico Fermi, Werner Heisenberg, and Wolfgang Pauli on an excursion boat on Lake Como during a summer physics conference, 1927.

8. The fifth Solvay conference, October 1927. Max Planck is the second from the left in the front row, and to his left are Marie Curie, Hendrik Lorentz, and Einstein. In the second row are Hendrik Kramers between Curie and Lorentz, then Paul Dirac to his left, and skipping one, Louis de Broglie, Max Born, and Bohr. In the third row, Ehrenfest is third from the left, Erwin Schrödinger at center, and Heisenberg and Pauli, third and fourth from the right.

9. Dirac and Heisenberg standing together, c. 1928.

10. Pauli and Ehrenfest sitting on the deck of the ferry taking them to Copenhagen, April 1930.

11. The 1930 Copenhagen meeting. Front row, left to right, are Oskar Klein, Bohr, Heisenberg, Pauli, George Gamow, Lev Landau, and Kramers. Felix Bloch is in the second row directly behind Gamow, and Rudolf Peierls is in the second row between Heisenberg and Pauli.

12. Bohr and Heisenberg skiing together in the early 1930s.

13. Landau, Gamow, and Edward Teller, with two of Bohr's children, in Copenhagen, early 1933.

14. The 1933 Copenhagen meeting. Front row, left to right, are Bohr, Dirac, Heisenberg, Ehrenfest, Delbrück, and Meitner. Carl Friedrich von Weiszäcker is in the second row directly behind Bohr; Peierls is at the right end of that row. Bloch is the second from the right in the third row, and Victor Weisskopf is at the right end of the fourth row.

15. Stockholm train station, December 1933. Right to left, Schrödinger, Heisenberg, Dirac, Dirac's mother, Schrödinger's wife, and Heisenberg's mother. Heisenberg has arrived with his mother to accept the 1932 Nobel Prize in Physics, awarded to him in 1933. Dirac, accompanied by his mother, and Schrödinger, with his wife, Anny, have come to accept the shared 1933 physics prize.

16. Bohr, Born, and Delbrück having a discussion during a break in the 1936 Copenhagen meeting.

17. Pauli and Bohr in the mid-1950s, studying a spinning top.

18. Niels and Margrethe Bohr surrounded by children and grandchildren, 1955. The occasion was Bohr's seventieth birthday.

19. Dirac and Richard Feynman discussing a physics problem at a meeting, 1962.

Heisenberg had shown that it is impossible to simultaneously measure p and q, momentum and position, with the precision needed to predict a particle's future trajectory. Some uncertainty, ultimately linked to the fact that Planck's constant is nonzero, is always left over.

Knowing that convincing the community was not going to be easy, Heisenberg also presented his conclusions in a nontechnical form to a German magazine. In the decades following the principle's enunciation, the notion that there is an intrinsic limit to what can be determined has been embraced as a metaphor in many fields, such as economics, where supply and demand clearly affect each other, creating uncertainty in a market situation. The concept has often proved useful in those areas, but one must remember that in physics the choice of variables is specified and the limit on simultaneous measurements is Planck's constant. In other fields both the choice and the limit are subjective. Nevertheless it remains the one concept of quantum mechanics that is universally alluded to, even if the details of the principle are glossed over. Ultimately its greatest use outside of physics may have been to focus attention on what happens through the act of making a measurement.

This is precisely the issue that concerned Bohr and the one that he felt Heisenberg had not settled adequately. While skiing in Norway, Bohr pondered the same question he had been thinking about for over a decade. What is the difference between the classical world and the quantum one? His correspondence principle had originally been the link between the two, subtly used to guide researchers toward quantum mechanics. Now, mathematical underpinnings in hand, he faced the final question: What are particles and what are waves? Schrödinger was emphasizing waves, and Heisenberg particles, but weren't they both right?

Bohr's answer was his complementarity principle, formulated in its definitive version during the summer of 1927. Once again Bohr's talent for joining together seemingly contradictory theoretical constructs came into play. The complementarity principle maintains that the concepts of wave and particle are simply attempts to describe a packet of energy using familiar terms. The packet is not *either* a wave *or* a particle. In a sense it is *both*, but the two identities cannot be seen simultaneously. Waves and particles are complementary ways to describe the results of a measurement. More accurately, the packet is *neither* a wave nor a particle. These are two ways we record the passage of a packet through our measuring apparatus. The act of measuring is the key, determining the identity we see in our experiments.

These concepts are not easy to understand. As Richard Feynman said in his lectures to California Institute of Technology students,

> Because atomic behavior is so unlike ordinary experience, it is very difficult to get used to and appears peculiar and mysterious to everyone, both to the novice and to the experienced physicist. Even the experts do not understand it the way they would like to, and it is perfectly reasonable that they should not, because all of direct, human experience and of human intuition applies to large objects. We know how large objects will act, but things on a small scale just do not act that way. So we have to learn about them in a sort of abstract or imaginative fashion and not by connection with our direct human experience.

Feynman then launches into an explanation of what happens when a beam of electrons is aimed at a barrier with two slits cut into it. Will they act like particles or like waves? A beam of electrons striking a partition with a double slit yields the characteristic wave interference pattern on a screen behind the slit. It will, however, only do so as long as we don't set up an apparatus to record which slit each electron went through. The identification of electrons as particles erases any possibility of detecting their wave nature; vice versa, a measurement of the packets as waves erases our ability to know them as particles.

The act of making a measurement and how we do it determine which form we see while precluding the possibility of seeing both forms simultaneously. But why can't such a concurrent measurement be made, thereby falsifying the notion of complementarity? This is where uncertainty and complementarity, Heisenberg and Bohr, go hand in hand. Complementarity would not be true if precise simultaneous measurements could be made, but they cannot because of Heisenberg's uncertainty principle.

Difficult adjustments in thinking were, and still are, demanded if one is to grasp the meaning of quantum mechanics. Born had said that one could only state relative probabilities of measurements, and Heisenberg had invalidated strict causality, asserting one could not determine present conditions with certainty. Now Bohr was maintaining that waves and particles are simply complementary ways of describing the results of experiments. Given the novelty and complexity of these ideas, it is not surprising that they were only reluctantly accepted at first, and not by everyone.

By the end of the summer of 1927, Bohr and Heisenberg, both ready to present their conclusions, were looking forward to the fifth Solvay conference. It was scheduled for October, in Brussels as usual. Solvay's convener, the wise elder statesman Hendrik Lorentz, picked the conference topic "Electrons and Photons" knowing that this choice would generate a much-anticipated debate on particles, waves, and the meaning of quantum mechanics.

Chapter 9

The King in Decline

MEPHISTOPHELES *(SINGS)*
A king there once was reigning,
Who had a goodly flea,
Him loved he without feigning,
As his own son were he!

Goethe, *Faust, Part I,* 1860–63

The Crucial Solvay Conference

BOTH THE TIMING of the Solvay conference and its topic could not have been more propitious. Almost a decade after the war's end, the world was entering a phase of conciliation and cooperation. By October 1925 the mood had advanced far enough for seven European nations to formulate a series of agreements to, among other things, guarantee the borders of Germany with France and Belgium and create arbitration processes to resolve potential disputes. Quickly approved, they came to be known as the Locarno Pact after the city where they originated. In the conciliatory mood that now existed, Germany was admitted in October 1926 to the League of Nations with a permanent seat on its council.

The postwar depression was replaced by ebullience and excitement over new art forms, ideas, and technologies. People admired the novelty of air flight, drove automobiles, talked on telephones, and flocked to theaters to see the first talking motion picture. A few months before the

Solvay conference a young American had become, at least for a while, the most famous person in the world. On May 20–21, 1927, Charles Lindbergh flew from New York to Paris in his *Spirit of Saint Louis*, a plane his biographer referred to as a "two-ton flying gas tank." Bouncing to a start, sometimes flying through fog only a few meters above the waves, he made the trip in a little over thirty-three hours. He took with him on his flight even less than Heisenberg had to Helgoland. When reporters observed he had only five sandwiches, he said with a succinctness worthy of Dirac, "If I get to Paris, I won't need any more. And if I don't get to Paris, I won't need any more either." Lindbergh was two months younger than Heisenberg.

In this exhilarating time, with stretching boundaries and challenging thoughts of all sorts, the world's leading experts on the subject of "Electrons and Photons" gathered in Brussels. Thanks to the "spirit of Locarno," Germans attended in force—or almost. Sommerfeld, who had proposed annexation of Belgium during the war, was not invited, even though, on the basis of merit, he should have been. But that was an anomaly. The twenty-eight physicists at the meeting included all the other major players in quantum mechanics' development: in order of age, Max Planck, Albert Einstein, Paul Ehrenfest, Max Born, Niels Bohr, Erwin Schrödinger, Louis de Broglie, Hendrik Kramers, Wolfgang Pauli, Werner Heisenberg, and Paul Dirac. It was the only time this whole group was together.

It turned out to be the most famous conference in physics' history, the one we physicists still talk about today. Signaling the completion of quantum mechanics' formative period, the conference marked the end of the revolution in quantum theory that had begun slightly more than two years earlier. With the Copenhagen interpretation, this greatest discovery of twentieth-century physics was now fully formed (although several other achievements could mark that terminus). The other reason the conference is famous is because it saw the beginning of the great debate between Einstein and Bohr about the meaning of quantum mechanics. As the novelist-scientist C. P. Snow said:

> No more profound intellectual debate has ever been conducted—and, since they were both men of the loftiest spirit, it was conducted with noble feelings on both sides. If two men are going to disagree, on the subject of most ultimate concern to them both, then that is the way to do it.

Up until then, these two great figures had disagreed on issues, such as the Bohr-Kramers-Slater proposal, but had eventually been able to reconcile their positions. This was no longer possible after Solvay. Nonetheless each one's affection for the other never wavered. More than four decades after their first meeting, Bohr reminisced about his old friend: "Einstein was so incredibly sweet. I want also to say that now, several years after Einstein's death, I can still see Einstein's smile before me, a very special smile, both knowing, humane and friendly."

Einstein was simply unwilling to accept the Copenhagen interpretation with its emphasis on how a measurement was performed and the underlying limits on obtainable answers dictated by Heisenberg's uncertainty principle. Some people think the reason for Einstein's refusal is that, absorbed in trying to generalize his magnificent ideas of space and time, he never gave his full attention to quantum theory, but this is far from the case. "I have thought a hundred times as much about the quantum problems as I have about the General Relativity Theory," he once said to a friend. No, the reason for his opposition to the accepted formulation of quantum mechanics is that it went against the grain of what he deeply believed to be the truth. Perhaps he was even right, although there is no evidence for this. The argument still goes on.

The Solvay conference's presentations were excellent, but the conversations about quantum mechanics between Bohr and Einstein, carried on in the hallways, during walks, and over meals, were the meeting's real heart. As they argued with each other, their friend Paul Ehrenfest was usually with them, the third party to the discussions. A letter he wrote back to his students in Leiden is probably the best record of what went on between Bohr and Einstein:

> Bohr towering completely over everybody. At first not understood at all, then step by step defeating everybody . . . Every night at 1 a.m. Bohr came into my room just to say ONE SINGLE WORD to me, until 3 a.m. It was delightful for me to be present during the conversations between Bohr and Einstein. Like a game of chess. Einstein all the time with new examples. In a certain sense a sort of Perpetuum Mobile of the second kind to break the UNCERTAINTY RELATION. Bohr from out of the philosophical smoke clouds constantly searching for the tools to crush one example after the other. Einstein like a jack-in-the box: jumping out fresh every morning. Oh, that was priceless. But I am almost without reservation pro Bohr and contra Einstein.

Ehrenfest then goes on to describe the details of the arguments, punctuating his narrative with an occasional "BRAVO BOHR!!!!!" Privately, he scolded Einstein for being so resistant, accusing him of beginning to sound like the conservatives who had once opposed the theory of relativity. Einstein wouldn't budge. The Old One had spoken to him, and he put more faith in Him than he did in either Bohr or Ehrenfest.

When Einstein talked of the Old One, he wasn't invoking a traditional divinity. His god was the god of Spinoza, the god of order in the universe. The notion that the observer necessarily affects the results of experiments was unacceptable to him. And yet he could not convince those who thought otherwise that his view was correct.

The 1927 Solvay conference was the watershed in quantum mechanics. At its beginning, only Bohr, Heisenberg, and Pauli were confident that a consistent formulation had been reached. Afterward, its general acceptance never faltered, even though some physicists, following Einstein, continue to have their doubts.

Thinking back on that conference and the ensuing years, more than a half century after its occurrence, Dirac, by then an old man, remembered what it had been like for him and offered his view of the future, including a few practical suggestions:

> In these discussions at the Solvay conference between Einstein and Bohr, I did not take much part. I listened to their arguments, but I did not join in them, essentially because I was not very much interested. I was more interested in getting the correct equations . . . it seems clear that quantum mechanics is not in its final form . . . I think it is very likely, or at any rate quite possible, that in the long run Einstein will turn out to be correct, even though for the time being physicists have to accept the Bohr probability interpretation, especially if they have examinations in front of them.

Now, almost eighty years after Solvay, the repeatedly verified Bohr interpretation still stands, as solid as ever, but still questioned, as it should be.

The conference was also important in ways that had nothing to do with the interpretation of quantum mechanics but everything to do with the careers of the young revolutionaries. The presence in Brussels of Heisenberg, Pauli, and Dirac marked their arrival as full-fledged members of the physics establishment. Despite their youth, they were now the intellectual peers of Bohr, Born, and Ehrenfest. Schrödinger, closer in age to the latter group than to the former, had in the meantime become enshrined as Planck's successor in Berlin.

Recognition came quickly and deservedly to the young trio as well. In December 1927 Pauli was given a full professorship at Zurich's polytechnic institute, the ETH (Eidgenössische Technische Hochschule). From then on he signed his papers as Wolfgang Pauli, dropping the "junior" he had heretofore used. Now he, not his father, was *the* Pauli. Dirac was made a fellow of Cambridge's St. John's College in 1927 and a fellow of the Royal Society in 1930. In 1932, when Sir Joseph Larmor stepped down from the Lucasian professorship, the Cambridge chair Newton had occupied, Dirac was appointed as his replacement, appropriately anointed as Britain's greatest theoretical physicist.

I suspect acknowledgment was particularly important for Heisenberg, the most ambitious of the three. A little more than a year after he had declined an associate professorship at Leipzig, Heisenberg was asked to became head of its Institute for Theoretical Physics. Leipzig was so eager to attract him that they agreed to give him an eight-month leave of absence in 1929 so he could travel around the world.

This quick recognition went a long way to ironing out the difficulties between Bohr and Heisenberg. Though their relationship had been strained to almost a breaking point in the beginning of 1927, their subsequent agreement on quantum mechanics' main features smoothed over the rift. By the autumn of 1928, after a brief visit to Copenhagen, Heisenberg wrote Bohr, "It made me so happy to see that once again we understood each other so well and that everything was once again as in the 'old days.'" Bohr replied, "Thanking you for the great pleasure you gave us all by your visit. Rarely have I felt myself in more sincere harmony with any other human being."

Einstein/the King in profile.

Einstein—The King

ALTHOUGH BOHR DISCUSSED his complementarity principle at the Solvay conference, he was even more reticent than usual in publishing its definitive version. Knowing its importance and sensing that the manuscript explaining it had to be clear, he asked Pauli for help in expressing his ideas concisely. They met several times over the next few months until Bohr's paper was deemed ready.

Two essentially identical versions appeared in journals simultaneously, a German account in *Naturwissenschaften* and an English one in *Nature*. Bohr's conclusions about the meaning of quantum mechanics were so provocative that *Nature* accompanied the article with a commentary in which its editors expressed their wish that this not be the final word on quantum mechanics since its conceptions could apparently not be "clothed in figurative language." Perhaps *Nature*'s editorial board hoped that causality would be restored and that one could determine when a particle is a particle and a wave is a wave. In a letter to Bohr, Pauli mocked their commentary with this paraphrase:

> We British physicists would be awfully pleased if in the future the point of view advocated in the following paper should turn out to be not true. Since however Mr. Bohr is a very nice man, such a pleasure would not be kind. Since moreover he is a famous physicist and more often right than wrong, there remains only a slight chance that our hopes will be fulfilled.

In spite of *Nature*'s ambivalent reception, most of the younger members of the physics community, as well as some of the elders, had come around to the Copenhagen interpretation of quantum mechanics by the end of 1928. As their views spread, books expounding the new approach appeared rapidly in German, French, English, and even Italian. But Einstein was still adamantly opposed to its concepts. Knowing that at least Schrödinger was in substantial agreement with his point of view, he wrote him, "The Heisenberg-Bohr tranquilizing philosophy is so delicately contrived that for the time being it provides a gentle pillow for the true believer from which he cannot very easily be aroused. So let him lie there."

Einstein's perceived obstinacy saddened some of his old friends. Max Born expressed the loss poignantly: "Many of us regard this as a tragedy—for him, as he gropes his way in loneliness, and for us, who miss our leader and standard-bearer." Others were more direct. Having exchanged letters with Einstein for almost a decade, Pauli now scolded him for an apparent lack of interest in the subject's development, writing him that "for the time being you don't want to hear anything more about quantum theory. I know that, but I'm very sorry to see it happen." Einstein was growing increasingly detached from the community of young physicists seeking to apply the new rules of quantum mechanics to a vast host of problems. He had embarked on another quest, the unification of gravitation with electromagnetism, and few were eager to follow him along that road.

But Pauli did not ignore either Einstein or his efforts. After reading Einstein's latest work in 1929, Pauli wrote him a scathing critique in December of that year. As I've stated earlier, one of the reasons physicists had such affection for the never-shy Pauli, even when his criticisms verged on insulting, was they knew his tone toward Einstein or Bohr was no different than it was toward them:

> I would like to add something about my opinion and that of a large part of the younger physicists about the physical side of this work of yours . . .
>
> I should congratulate you (or should I rather send condolences?) that you have switched to pure mathematics.

Pauli then goes on to wish Einstein's newest effort a speedy death.

By return mail, Einstein rebuked Pauli: "I found your letter very amusing, but your opinion seems rather superficial. One should only write the way you do if one is sure about his point of view with respect to the unity of the forces of Nature." He then urged Pauli to take a fresh look at the problem, telling him to act as if he had just dropped on Earth from the moon: "Study the problem for a few months and then tell me what you think about it." Pauli never did. Einstein, though unwilling to abandon his quest, was certainly willing to admit his mistakes. A January 1932 letter to Pauli begins, "Dear Pauli: You were right, you *Spitzbube*" (rascal).

Pauli was wrong, however, on one count. Einstein was continuing to think about quantum mechanics. When the sixth Solvay conference was held in October 1930, the announced topic was magnetism, but once again the meeting's most famous exchange was Einstein's challenge to the Copenhagen interpretation and Bohr's response to it. Einstein proposed the following setup: A box contains particles and a clock; it also has a shutter that opens and closes at a specified time, allowing only one particle to come out. The box is weighed before and after the shutter opens. Since $E = mc^2$, the difference in weight is a measure of the particle's energy. The instant of shutter opening sets the time when the particle leaves the box. The process appeared straightforward, but according to the uncertainty principle an exact measurement of a particle's energy cannot be made at a prescribed time.

Bohr, at first deeply troubled by Einstein's argument, had an answer ready by the next day. Einstein had neglected to take into account that weighing a box involves a slight movement in the earth's gravitational field, in turn leading to a small uncertainty in the determination of mass and hence of energy. Furthermore, as Einstein himself had shown years earlier, the running of a clock changes with its position in a gravitational field; in other words, there is an uncertainty in time as well. Bohr then showed that the product of the uncertainties satisfies Heisenberg's uncertainty principle!

He had turned Einstein's seeming paradox into a splendid confirmation of the uncertainty principle by an ingenious application of his adversary's own theory of relativity. Einstein was impressed by Bohr's refutation of the carefully constructed counterexample to the Copenhagen interpretation, realizing also that it had been found so extraordinarily quickly because Bohr's faith was unshakable. But Einstein's faith was equally unshakable. The battle was lost, not the war.

This was the final test for most physicists, though questions about the Copenhagen interpretation have continued to this day. It was also the last public discussion between Bohr and Einstein. By the time of the seventh Solvay conference, October 1933, Hitler had taken control of Germany, and Einstein had left for the United States, never to return to Europe again. But 1930 wasn't the end of the great debate between Bohr and Einstein. In the next two decades they often met at the Institute for Advanced Study in Princeton, New Jersey, their discussions continuing in private until Einstein's death in 1955.

But these discussions of Bohr and Einstein also seemed increasingly irrelevant to the physicists who came of age after the quantum mechanics revolution, the under-thirty crowd in 1932 Copenhagen. They simply ignored Einstein's continuing contributions to physics. Though they revered him, they felt no need to pay attention to his work, since it did not seem to have any impact on theirs, nor did it indicate new directions they wished to explore. By then in his fifties, Einstein was twice their age, perceived by them as no longer involved in the up-to-date problems that commanded their attention.

However, the debate about the interpretation of quantum mechanics that Bohr and Einstein started has never died. Questions arose again in 1935 around the so-called Einstein-Podolsky-Rosen paradox. The late John Bell, an extraordinary theoretical physicist, rekindled the debate again in the 1960s. Even a new and lively field, commonly known as quantum computing, has emerged and thrived from these notions combined with technological developments, demonstrating yet again the magnitude of the revolution.

Chapter 10

The Great Synthesis

FAUST
That I the force may recognize
That binds creation's inmost energies;
Her vital powers, her embryo seeds survey,
And fling the trade in empty words away.

Goethe, *Faust, Part I*, 29–32

Dirac's Equation

"I WAS MORE INTERESTED in getting the correct equations" was Dirac's reaction to the Bohr-Einstein debate at the 1927 Solvay conference. A few months later, Dirac did get the correct equations and in doing so he figuratively joined Bohr's domain of quantum mechanics to Einstein's of relativity.

Both Heisenberg and Schrödinger used the framework of classical mechanics for their examination of the quantum world. They knew that adding the constraints of relativity would introduce an extra layer of complexity, an unnecessary one for the problems they were interested in. However, physicists studying the quantum world knew that eventually they would need a theory that described electrons moving at any speed up to relativity's ultimate limit, that of light. They presumed it would be a generalization of Schrödinger's equation with relativity theory built into it. This was the equation Dirac found at the end of 1927.

Its predictions for the behavior of electrons in electric and magnetic fields agreed spectacularly well with experimental data, so much so that the equation was immediately hailed as a breakthrough. Linking his approach to Schrödinger's, Dirac said,

> Schrödinger and I both had a very strong appreciation of mathematical beauty, and this appreciation of mathematical beauty dominated all our work. It was sort of an act of faith with us that any equations which describe fundamental laws of Nature must have great mathematical beauty in them. It was like a religion with us. It was a very profitable religion to hold, and can be considered the basis of much of our success.

The equation's beauty even stunned Dirac. In a revelatory lecture he gave late in life, he explained the apprehension he had felt fifty years earlier upon discovering it and why he had stopped on the threshold of finding an exact solution for its prediction of electron motion. He could have done so without too much effort, but the nightmare of having to discard his findings held him back.

> I was really scared to do so. I was afraid that, in higher approximations, the results might not come out right and I was so happy to have a theory that was correct in the first approximation that I wanted to consolidate this success by publishing it in that form, without going on to risk a failure in the higher approximations . . . The originator of a new idea is always rather scared that some development may happen which will kill it, while an independent person can proceed without this fear, and can venture more boldly into new domains.

Though its exact solution, found by others, agreed spectacularly well with experiment, the equation did have one deeply troubling aspect. The compact form in which it is usually presented leads physicists to speak of it as a single equation, but it actually is four coupled equations, with four solutions. Two of the solutions were evident, corresponding to the motion of the electron with spin either up or down. The other two were seemingly nonsensical: a spinning electron with negative energy. What could it possibly mean that an electron had negative energy? If $E = mc^2$, this seemed to imply that electrons could have negative mass. But there are no such objects. The situation was particularly worrisome because quantum theory indicated that electrons should be able to jump from having positive to having negative energy.

The existence of the equation's four solutions would have ruled out further consideration were it not for its striking agreement with experiment in other aspects and its intrinsic elegance. As it turned out, the equation was correct; it simply took a few years to understand why there were four solutions instead of two, an important story we will come to in the next chapter.

Dirac's equation put into place the last of quantum theory's building blocks. By 1930 the physicists who had shaped the revolution's beginnings had already received Nobel Prizes. Planck, Einstein, Bohr, and de Broglie had made the trip to Stockholm. It was now time to celebrate the new generation, but who should go first, Heisenberg or Schrödinger? Heisenberg's matrix mechanics came earlier, but Schrödinger's wave mechanics was still more influential. And what about Pauli, Born, Jordan, and Dirac? Pauli wrote the Nobel committee recommending that the award go to Heisenberg and then to Schrodinger, while Einstein suggested first Schrödinger and then Heisenberg. Bohr opted for both together. Seemingly unable to decide, the Stockholm committee did not award the physics prize to anyone in 1931. Still uncertain in 1932, they passed once more. Finally, in 1933, they managed to reach a decision: the 1932 prize was awarded to Heisenberg, and the 1933 prize was shared by Dirac and Schrödinger. The choice left some hard feelings; Heisenberg wrote an apologetic letter to Born and a second letter to Bohr, saying, "Schrödinger and Dirac deserve an entire prize at least as much as I do, and I would have gladly shared with Born."

Since the 1932 and 1933 prizes were awarded together, we have a December 1933 photograph at the Stockholm train station of a smiling

Schrödinger, wearing a bowtie and knickers, with his wife, Annemarie (Anny), while the more serious Dirac and Heisenberg, looking like dutiful children, are accompanied by their mothers (see photograph 15). *Knabenphysik* had made its way to Stockholm.

How Max Delbrück Joined *Knabenphysik*

IN THE LATE 1920s and early 1930s many bright students were drawn to physics by the exciting development of quantum mechanics. The more successful ones had brilliant careers; several won Nobel Prizes and were deeply influential. Yet despite being only a few years younger than Heisenberg, Pauli, and Dirac, they were always aware that a gulf divided them from the founders of the quantum mechanics revolution.

Max Delbrück, a member of this new generation, was twenty years old when, in September 1926, he enrolled at the university in Göttingen, transferring from his native Berlin. Arriving in the small German town, he found the controversy between matrix mechanics and wave mechanics at its height. Delbrück, not yet an active researcher, could at first only admire from a distance the emerging group devoting itself to applications of quantum mechanics. Most of its members were under twenty-five, swelling rapidly the ranks of *Knabenphysik*. Delbrück thought that with a little luck he would quickly join them.

Pascual Jordan, twenty-four years old, a charter member of *Knabenphysik*, had written key papers with Born and Heisenberg; he now had been appointed *Privatdozent* (lecturer) in Göttingen and was beginning to teach.

Eugene Wigner, twenty-three, a Hungarian, would one day introduce new branches of mathematics into quantum theory, work for which he eventually received the Nobel Prize in Physics in 1963. He too came to Göttingen that fall. In his native Budapest, Wigner had hesitated between mathematics and physics but thought the latter was a more promising prospect, since a high school classmate was so much more talented than he in mathematics. What he couldn't have known was that his classmate,

John von Neumann, would go on to be one of the century's great mathematicians, as well the developer of the first digital electronic computer and the inventor of game theory in economics. A true universal genius, von Neumann also had a strong interest in quantum mechanics and was already doing much to put the new field on solid mathematical footing.

Then there was the brilliant young American who arrived the same time as Delbrück and immediately started writing a thesis with Max Born. J. Robert Oppenheimer seemed to understand everything at lightning speed. In one year "Oppie" and Born established a basis for a quantum mechanical treatment of molecules.

In a little piece entitled "Homo Scientificus According to Beckett" written almost fifty years later, perhaps as an homage to Samuel Beckett, who received the Nobel Prize in Literature the same year Delbrück was honored in physiology or medicine, the physicist-biologist remembered what it was like in Göttingen at the time:

> I found out at an early age that science is a haven for the timid, the freaks, the misfits. That is more true perhaps for the past than for now. If you were a student in Göttingen in the 1920s and went to the seminar Structure of Matter which was under the joint auspices of David Hilbert and Max Born, you could well imagine that you were in a madhouse as you walked in. Every one of the persons there was obviously some kind of a severe case. The least you could do was put on some kind of stutter. Robert Oppenheimer as a graduate student found it expedient to develop a very elegant kind of stutter, the njum-njum-njum technique. Thus, if you were an oddball, you felt right at home.

Oppenheimer's stutter, which disappeared after he left Göttingen, must have made a big impression on Delbrück, as evidenced by a scene in the "Copenhagen Faust." When Mephisto enters an American speakeasy to sing the song about the King with the pet flea (more about this later), the stage directions indicate that a young man named Oppie responds to the devil's query

American physicists sitting sadly at the bar of an Ann Arbor speakeasy as Pauli/ Mephisto enters.

> *Can no one laugh? Will no one drink?*
> *I'll teach you Physics in a wink. . . .*

with a stutter, saying Njum! Njum! before speaking the line

> *Your fault! You've brought no single word of cheer—*

Göttingen physicists seemed so bright to Delbrück, if perhaps a bit strange. Furthermore he could not console himself by saying they were all older than he was since Maria Goeppert, who later shared the Nobel Prize with Wigner, was his own age, and brilliant Victor (Viki) Weisskopf (like Meitner, Ehrenfest, and Pauli, yet another Jew from Vienna) was younger than he and yet seemed to be absorbing the new quantum mechanics theory faster than he could. No matter! Weisskopf became his lifelong friend, their paths across continents continually intersecting. In any case, he thought that even if he wasn't the smartest of the group, quantum mechanics was so intriguing and so rich that there would certainly be something for him to do as well.

The good news was that Max Born agreed to supervise his thesis. But Born, who had arrived as a professor at Göttingen in 1921, was buckling by 1928 under the strain of understanding the rapid scientific changes and of being the warden of the "madhouse." Now forty-five years old, he felt unable to maintain his pace. As he later remembered

> It was very difficult for me, a man of years, to keep up with the young ones. I made strenuous efforts, but these led to a nervous breakdown (1928) which forced me to interrupt teaching and research for about a year and to go more slowly afterwards.

Born tried to help Delbrück but could not be of much assistance. Fortunately yet another talented young physicist arrived in Göttingen at that moment, filling the envied position as Born's assistant, the successor to the likes of Pauli and Heisenberg. Out of necessity he assumed many of Born's responsibilities and suggested a thesis topic to Delbrück.

Walter Heitler, the assistant, had shown the year before how quantum mechanics explains the binding between two hydrogen atoms to form a hydrogen molecule. It was an important calculation, many say the true beginning of quantum chemistry and at least in part the reason for Dirac's previously quoted 1929 remark: "The underlying physical laws necessary for the mathematical theory of a large part of physics and the whole of chemistry are thus completely known."

Heitler suggested that Delbrück follow up this calculation by examining the binding between two lithium atoms. Helium, number 2 in the periodic table, was known to be very different from hydrogen, but lithium, number 3, was expected to have a similar behavior. Of course the calculation would be more complicated than his, but he would be able to help Delbrück when he ran into difficulties.

This is exactly the kind of thesis topic one assigns a bright young student: long and perhaps tedious, but providing an answer that will be relevant whatever the outcome, almost sure to be enough for a doctorate as long as the student is not scooped. Because the world of physics at that time was still small, Heitler would have known if anybody else was studying lithium-lithium binding; being scooped was unlikely even if Delbrück wasn't able to do the work as fast as a more seasoned physicist.

The calculation did turn out to be tedious. No new ideas were involved, and the result, as Delbrück put it in his usual understated way, was "acceptable but rather dull." After it was finished and the results submitted to a journal, he accepted a nine-month position in the University of Bristol physics department. By 1930 Delbrück was back in Germany, the recent winner of a Rockefeller Foundation fellowship, an award that would support him for a year's research any place he wished. He was now a full-fledged member of *Knabenphysik*.

With his fellowship in hand, Delbrück began thinking about his next move. There were so many interesting problems in quantum mechanics to work on. What should he do?

Physics Begins to Split

THOUGH NEITHER DELBRÜCK nor his cohort realized it, physics at the end of the 1920s was taking a step toward increased specialization. This happens in any profession when the training period stretches and the amount of knowledge needed for progress increases. Doctors, who not too long ago simply treated sick people, are now divided into internists, surgeons, ophthalmologists, dermatologists, psychiatrists, and the many other specialties we consult, these in turn branching off into further subdisciplines.

A division much like this occurred in physics with the development of quantum mechanics. Before the 1920s physicists worked at a great variety of problems in many areas, but such breadth became less frequent afterward. Quantum mechanics provided tools that were so powerful that learning the subject became de rigueur, but applying these new techniques required further training, so the profession began to separate into subspecialties. Physicists soon started defining themselves as condensed matter physicists, atomic physicists, nuclear physicists, or astrophysicists, to name a few. They in turn branched off again, leading to almost as many specialists as the medical profession.

For a while individuals crossed these boundaries, but there was little of that by the 1960s, the time I joined the physics department at a large American research university. We shared the teaching of the basic undergraduate courses and tried to speak with one voice to the dean, but we met separately to plan our research activities and to request funding, each specialty having by then progressed to the point of having its own journals—and its own meetings, such as the gathering of neutrino experts described in this book's opening.

We looked back with nostalgia to the not so remote past, still within our own lifetime, when one physicist could bridge all fields of theoretical physics. Even more rare, some had then seemed capable of doing significant work in both experiment and theory, almost unthinkable by the time I became a physicist. As my uncle Emilio wrote of his mentor Enrico Fermi, who died in 1954,

> Thus disappeared the last physicist who dominated the whole field both in theory and in experiment. It is doubtful whether, with increasing specialization, we will ever again see such a universal excellence.

Of course the other side of the coin is the appreciation of the great progress made through this increased specialization. It has come at a cost, but a cost we are happy to bear when gazing at the wonders it has produced. Nor can we ever be quite sure where the next wonders will come from, because it is in their nature to surprise and amaze us.

Never knowing a priori what will be required next or who is more likely to advance the field, we also need contributors with varied talents, different tastes, and different styles. It is doubtful that Pauli could have found the relativistic equation that Dirac discovered because Dirac's "religion" of looking for equations with "great mathematical beauty" was one that Pauli had trouble embracing. Though Pauli was a skilled mathematician, he was far more wedded to being guided by experimental data than, as Dirac put it, "playing with equations." On the other hand it is extremely unlikely that Dirac, with his emphasis on mathematical beauty, would have performed the careful analysis of experiments that led Pauli to formulate his exclusion principle or would have suggested that experimentalists search for neutrinos. And probably neither of them could have arrived at Heisenberg's uncertainty principle. Each of these discoveries bears the stamp of its creator's attitude toward physics.

The debate on whether to be guided by experiment or by mathematical elegance remains a classic physics dilemma, probably most strongly felt by theoreticians, the practitioners working closest to the line between the two. Not surprisingly, a particular theory's elegance is usually most striking to the one who loves it best, its designer.

Of course experiments can give wrong results, so it may be a good idea for a theorist to wait until the experiment has been confirmed before throwing away his or her theory. This was the case with the celtium-hafnium controversy, when the French results seemed to contradict Bohr's predictions. So far we have discussed correct theories preceding and following correct and incorrect experiments. But history tends to tell the story of the victors, and this is one situation where there were many more losers than winners. Wrong theories outnumbered correct ones by a wide margin, some of them stimulated by wrong experiments. Wrong theories and wrong experiments are a bad combination; two wrongs do not make a right. And all scientists agree that a theory must be discarded, no matter how beautiful it may appear, if its predictions contradict definitive experimental data.

There inevitably are shades of gray that cause the creator to hesitate in his or her judgment. To make matters even more complicated, sometimes a theory has not reached the point of being directly testable by experiments. In that case there is no risk of having the creation annihilated by contradictory facts. Then criteria of mathematical elegance may dominate, as in the case today of superstring theory. Advocates emphasize the theory's inner consistency, its ability to provide a unified picture encompassing all interactions, a "theory of everything" as its proponents sometimes call it. Its critics cite the lack of confirming experimental evidence.

In yet another common science scenario, a framework may be erected that explains previously mysterious experimental data, but with inner contradictions implying it only provides a provisional answer. As was the case with the Bohr model of the atom, this can be a very hopeful and stimulating state of affairs. Enough of that model's predictions were so strikingly verified that most scientists believed it was the first step toward a definitive correct theory. And that is what happened.

Sometimes there is room for subtle modifications, a moment when an old theory's basic features are shown to be only an approximation to a more general structure. Newton's theory of gravity remains the quintessential example. Though only an inexact solution to the equations of Einstein's general theory of relativity, it is more than adequate for describing planetary orbits unless you are concerned with such fine points as an apparent discrepancy in the precession of Mercury's perihelion by a few arcseconds per century.

The removal of this discrepancy is what convinced Einstein of the essential correctness of his theory years before the measurement of the deflection of light in a gravitational field brought it to general attention. In the words of Einstein's friend and biographer, Abraham Pais, the realization of this agreement was "the strongest emotional experience in Einstein's scientific life, perhaps in all his life. Nature had spoken to him." Perhaps this was the beginning of his confidence in the Old One.

Dirac's feelings when he discovered his equation were probably similar to Einstein's in 1916, though the confirmation for Dirac was not as dramatic as for Einstein, nor does one envision Dirac writing to a friend, as Einstein did to Ehrenfest in 1916, that "for a few days I was beside myself with joyous excitement." Both Dirac's and Einstein's equations have mathematical elegance to the nth degree. Intrinsically simple and yet strikingly novel, they led to new fields of inquiry in both mathematics and physics. Not designed to explain experimental observations, they did just that in a way their architects had not anticipated. Unable to be altered or adjusted, each is, as it were, chiseled in stone, fit for placement on its creator's headstone. Indeed, the Dirac equation is inscribed on Dirac's memorial in Westminster Abbey.

Dirac once talked about when "a person first gets a new idea and he wonders very much whether this idea will be right or wrong. He is very anxious about it and any feature in the new idea which differs from the old established ideas is a source of anxiety to him." The genre is different, but the feelings are the same as those in any creative endeavor, anxiety when ideas appear and depression in fallow times. Nor are the fears and anxieties I have been describing special to great physicists. We all have to manage them, if only by finding ways to deny them. My own peculiar technique of dealing with what I hope is a good idea is not to work on it for a few days. I want to savor the pleasure of having had the thought before beginning the inevitable and unfortunately usually successful task of finding the error in my thinking. On the other hand, I don't want to wait too long lest my hopes rise too high and the disappointment be too great; hope and excitement come together with anxiety. Such are the games one plays with oneself in this profession.

However troublesome having ideas may be, not having them is more painful, but even that is not the worst that can happen. In *Knabenphysik* the young may and often do simply speed away. It then becomes more like a race, say, a day in the Tour de France when the leaders are engaged in an Alpine breakaway from the pack, the *peloton*. Some followers struggle just to stay with the *peloton*, but then slowly even that becomes too hard. Nobody put it more poignantly than Ehrenfest in a 1931 letter to Bohr:

> I have completely lost contact with theoretical physics. I cannot read anything more and feel myself incompetent to have even the modest grasp about what makes sense in the flood of articles and books. Perhaps I cannot be helped any more.

Delbrück's Choices

As Delbrück and his cohorts began to make their choices of what problems to work on, their dominant thought was not of specialties, styles, or their own aging but rather how best to set out on their personal adventures in research.

Delbrück's expertise, acquired from his Göttingen thesis work and his time in Bristol, was in the quantum mechanical explanation of how atoms bind to form molecules, but he now wanted to examine more speculative questions, ones more likely to lead to dramatic new results. The two areas that seemed most exciting were the atomic nucleus's mysteries and the joining of relativity to quantum mechanics.

The former was particularly promising. Until the beginning of the twentieth century, scientists had not even been sure that atoms were real, and yet within the next ten years their existence had been confirmed, and Rutherford had even shown what they consist of: a minute nucleus surrounded by orbiting electrons. Bohr had then proposed a groundbreaking model of atomic electrons' motions, and by the late 1920s quantum mechanics had explained in glorious detail their interactions with those minute nuclei. But the inner structure of those nuclei was still a mystery.

The first window on the atomic nucleus had opened in March 1896 with Henri Becquerel's accidental discovery of a persistent strong radiation emitted by salts containing uranium. Two years later, in 1898, Ernest Rutherford, then twenty-seven, realized that the rays contained two distinct components. He gave them the names *alpha* and *beta*. Whereas beta rays were quickly shown to be streams of the very recently discovered electrons, the identity of alpha rays remained unknown for several years. Finally, in 1908, Rutherford proved that their constituents were helium atoms stripped of their negative charges; in other words the alpha particles that made up an alpha ray were simply helium nuclei.

A few years after Rutherford made his initial discovery, a third type of ray was found exiting from radioactive substances; in keeping with the

Greek alphabet, it was called a *gamma* ray. Later shown to be photons, quanta of electromagnetic radiation, gamma rays carried no electric charge.

Rutherford had decided early on to focus his attention on alpha rays. They were his preferred tools for discovery, and he was the world's recognized expert on their behavior. Two types of alpha ray experiments were commonly performed, both of which had been pioneered by Rutherford. The first was to study the details of the rays' emission, the key question being, How could some nuclei discharge helium nuclei from their interiors? The second was to bombard atomic targets with beams of alpha rays, examining the results from the collisions. This is how Rutherford's Manchester laboratory showed that nuclei exist. The researchers observed that alpha rays, usually unscathed when passing through a thin gold foil, were occasionally scattered at large angles and sometimes even rebounded almost directly backward. Rutherford inferred that this was due to the presence of something very small and very massive inside the gold atoms. He gave that something the name *nucleus*.

In the closing years of World War I, Rutherford, still in Manchester, performed his last great solo experiment. In that effort he showed that bombarding a nitrogen atom with alpha particles sometimes produces a hydrogen atom together with a form of oxygen. This was the initial case of induced nuclear transmutation, the first time an experiment had succeeded in changing one element into another. As Rutherford said at the time, "The results as a whole suggest that if alpha particles—or similar projectiles—of still greater energy were available for experiment, we might expect to break down the structure of many of the lighter atoms."

When Rutherford had discovered alpha and beta rays, he was a researcher in Cambridge, away from his native New Zealand for less than three years. In 1928, thirty years after his entry onto the world physics scene, Rutherford was the world's most influential experimental physicist, Sir Ernest, soon to be Lord Rutherford, Nobel Prize winner, Cavendish professor at Cambridge, and president of the Royal Society. Aging but still energetic, he had become more of a guide and facilitator for younger physicists than a hands-on experimentalist. That didn't mean, however, that his interest in physics had waned or that he didn't have strong preferences. While other laboratories might focus on experiments that investigated the behavior of atomic electrons, Rutherford felt the Cavendish should be the premier discoverer of the nucleus's secrets.

Though a great deal was already known about alpha, beta, and gamma rays, the mechanism or mechanisms by which nuclei emitted them were still a mystery in 1928. Quantum mechanics, which had revolutionized the understanding of electrons' motion about the nucleus, had still very little to say about the atom's core. Sir Ernest had accordingly largely ignored its development. He had so far not needed to learn the subject to deal with the nuclear physics questions that concerned him, and in any case he was not very interested in what he perceived to be abstract mathematics. As he used to say about theorists, "They play games with their symbols, but we in the Cavendish turn out the real facts of nature."

Delbrück thought that this was a very promising state of affairs. Here was nuclear physics, a new field containing important problems. Furthermore it might suit him well. He had never been as strong a mathematician as many of the group he had met in Göttingen and was therefore more likely to meet success if experiment, not mathematical elegance, was his guide. His style would be Bohr's and Pauli's, not Dirac's and Heisenberg's. Besides, even if Rutherford didn't think so, quantum mechanics was almost sure to eventually play a vital role in understanding the key features of the atom's core, the "real facts of nature."

Delbrück also suspected another revolution in physics might take place, perhaps with notions as unimagined as quantum mechanics had been a few years earlier. Nobody had initially suggested that grand new principles would be necessary to explain what occurs on the scale of the atom, a thousand times smaller than what could be seen with an optical microscope, but they had been. What surprises might now be waiting for physicists as they explored the domain of the nucleus, a hundred thousand times smaller than the atom as a whole?

Evidence that new forces and new principles were needed was already obvious from examining helium, the number 2 element in the periodic table. Rutherford had coined the term *proton* for the hydrogen atom's nucleus, the smallest known positive electrical charge. Since a helium nucleus was known to have twice as much positive charge as hydrogen, it seemed sensible to think of it as two protons, tightly bound to each other. But electrical charges of the same type (in this case, positive) repel each other with a force that becomes increasingly stronger as the distance between the charges shrinks. There was, however, no evidence of the expected repulsion between the two protons in the tiny helium nucleus. To the contrary, they clung tightly together.

Furthermore, a helium nucleus ought to have twice a hydrogen nucleus's mass if it is simply made of two protons, but its mass was instead known to be four times as large as a proton's. The answer to that riddle might be the presence inside helium nuclei of tightly bound electron-proton pairs. These would have no electrical charge and a mass approximately equal to that of the proton, since the electron's mass is negligibly small compared with a proton's. Add two of them to two protons and, voilà, you have the necessary ballast to form a helium nucleus. But it wasn't quite so simple. There had to be more than that to the solution, because why were electrons sometimes inside the nucleus and other times outside? Also, contradictions with Heisenberg's uncertainty principle arose from confining electrons inside the small space of a nucleus. Solving one puzzle seemed to produce two others. Was this progress? Might the rays emitted by nuclei carry some secret clues the physicists had not yet deciphered?

Delbrück knew that new experimental data and new ideas would be needed. But he also realized that, just as Bohr had done in his formulation of an atomic model, the old data would have to be understood as well. There certainly was a good deal of it, thirty years' worth. The existence of these great puzzles seemed promising to the young physicist. Maybe he and his friends would be the next generation's revolutionaries. At the very least they would be participating in a new and exciting intellectual adventure.

Chapter 11

Conservation of Energy

FAUST
What he can grasp, that only knoweth he.
So let him roam adown earth's fleeting day;
If spirits haunt, let him pursue his way;
In joy or torment ever onward stride,
Through every moment still unsatisfied!

Goethe, *Faust, Part II,* act 5, 407–11

The Mysteries of the Nucleus

SITTING IN HER LABORATORY one morning in 1928, Lise Meitner was thinking about the results recently obtained by Charles Ellis's group, her major rivals in the study of beta rays. She found their conclusions hard to accept, even though they seemed to be the outcome of long, painstaking research. She also trusted data from the Cavendish. Rutherford's laboratory, where Ellis had conducted his research, stood behind all its findings and rarely made mistakes. If these latest findings were correct, it meant that the puzzle about beta rays' behavior was, if anything, more challenging.

Even though Meitner had been studying these rays for almost a quarter century and was one of the world's experts in analyzing them, their strangeness continued to surprise her. Fortunately by 1928 she could devote herself to her investigations without the worries that had until recently afflicted her. Her academic position was now solid, and her private life was no longer marked by poverty. She had previously been living in a furnished room, often depending on friends for food. Writing to her widowed mother in 1923, she said, "You shouldn't send so much. I still have coffee, which I drink on Sundays." But soon thereafter both the German economy and her personal situation improved rapidly.

Like her colleague Otto Hahn, Lise Meitner had now become a professor as well as the head of one of the branches of Berlin's Kaiser Wilhelm Institute for Chemistry. Her expertise in physics and Hahn's in chemistry, with overlapping interests, had made them a formidable combination. She acquired a permanent assistant, began to have students, and visitors came to work with her. Finally, she now also managed to have her own apartment.

Meitner thought beta ray emissions from nuclei had very curious characteristics. Rutherford had initially assumed they originated as orbiting electrons, expelled for some reason from their presumably stable ambience. In 1913 Bohr showed him this was not the case. Whatever it was that produced alpha, beta, and gamma rays, they all came straight from the nucleus, not from the atom's outer sphere. But there was still a significant, puzzling difference. Alpha and gamma rays carried off, as expected, energies equal to the difference of the parent nucleus's energy before and after the transition, while beta rays did not. This seemed to imply that either energy wasn't conserved during their emission or some radiation, exiting simultaneously from the nucleus, had been missed. Both possibilities seemed unlikely.

The beta ray problem had surfaced in 1914 through the efforts of James Chadwick, then twenty-three years old, the son of a Manchester cotton-mill worker. Winning a scholarship to attend university there, he had been inspired by Rutherford to study physics. Having graduated with honors and gone abroad on a postgraduate fellowship, he was put to work examining beta rays in the Berlin laboratory of Hans Geiger, an old collaborator of Rutherford's.

Unfortunately, right after obtaining preliminary but significant results, Chadwick and five thousand other British civilians found themselves trapped behind enemy lines by the sudden start of World War I. They were imprisoned in an abandoned racetrack near Berlin, six prisoners to a horse stall. From his childhood Chadwick was accustomed to hardship, but this was like nothing he had ever experienced. With little heat and not enough to eat, the four years nearly killed him. Nevertheless he managed, thanks to Geiger, to obtain a some extra food and a few rudimentary pieces of equipment. Trying to keep his spirits up, he even did a bit of physics research, drafting as an assistant a young officer named Charles Ellis.

With the war over, Chadwick returned to Manchester; when Rutherford moved to Cambridge, he followed him there, quickly becoming his right-hand man, the person in charge of day-to-day operations at the Cavendish Laboratory. Ellis, who had originally planned an army career, had enjoyed his work with Chadwick—admittedly under difficult circumstances—and changed direction. He joined Chadwick and Rutherford at the Cavendish, obtained a doctorate, and soon was the laboratory's expert on beta rays.

It was Ellis's work that Meitner was puzzling over in 1928. The more she thought about it, the more Meitner realized that she would have to repeat the Cambridge experiments even if it meant months of work with relatively little glory at the end. A second opinion was going to be necessary, no matter how reliable Ellis's results were. There was always some possibility of an error in such research, and the standard operating procedure for an important experiment was and still is to have it repeated in another laboratory. If the results agree, the community can proceed with confidence.

While Meitner was planning her experiment, Rutherford was entering his fourth decade of working with alpha rays. Though he was no theorist and hadn't learned how to calculate using quantum mechanics, Rutherford had a good intuitive sense of the functioning of atoms and nuclei. He would begin thinking about a problem by forming for himself a simple picture of how a scattering or decay might take place, without invoking too much mathematical detail. Then he would use this picture in the planning and interpretation of experiments. This was, after all, the approach that had led him to discovering the atomic nucleus.

Following this method, Rutherford tried to imagine a model that could explain alpha ray production from radioactive nuclei such as uranium's. He envisioned a scheme in which positively charged alpha particles, each one electrically neutralized by two electrons, are located in the immediate neighborhood of a massive nucleus. At a certain point, for reasons that remained unclear, the two electrons start drawing the alpha particle away from the core, much like two tugboats pulling a liner out of the harbor. On getting to open water, the tugboats disengage and retreat back, leaving the liner (in this case, the alpha particle) to steam onward. The article describing this model appeared in a 1927 issue of *Philosophical Magazine*.

Given the author, it was respectfully read, but it contained many puzzling features. What made the *tugboats* leave the harbor, and why did they come back? The model did not even fit the data very well.

The Barrier Is Too High

A TWENTY-FOUR-YEAR-OLD Russian named George Gamow, lounging in the Göttingen library, was the first to glimpse the correct explanation. It was early June 1928. Gamow had arrived in Göttingen only a few days earlier from Leningrad, as Saint Petersburg was now called. One physicist who met him in Germany recalled what Gamow was like in those days:

> I shall never forget the first time he appeared in Göttingen—how could anyone who has ever met Gamow forget his first meeting with him—a Slav giant, fair-haired and speaking a very picturesque German, in fact he was picturesque in everything, even in his physics.

Since no senior Leningrad physics professor had learned quantum mechanics, Gamow had had no chance to obtain formal training in the subject, but he and two other students had formed a group that studied together the papers coming from the principal physics research centers. In the late spring of 1928, his professors tried to provide encouragement by arranging a three-month fellowship for Gamow in Göttingen. That explains why he was in the library on the day he had his remarkable insight.

Picking up the recent issue of *Philosophical Magazine*, Gamow saw Rutherford's paper on alpha particle emission. In his own words,

> Before I closed the magazine, I knew what actually happens in this case. It was a typical phenomenon which would be impossible in classical Newtonian mechanics, but was in fact to be expected in the case of the new wave mechanics. In wave mechanics there are no impenetrable barriers.

The phenomenon Gamow was referring to is sometimes called barrier penetration. Whether by that name or the alternative one of quantum tunneling, you can visualize how it operates if you imagine alpha particles held in the nucleus by a force, one whose details need not be specified. The force creates a barrier that traps the particles; it is a kind of prison. The rules of classical mechanics dictate that the prisoner can never escape, but quantum mechanics allows a surprising new freedom. As with any jail, the more massive the barrier, whether higher, wider, or both, the harder it is to tunnel through, but the chance of doing so is never zero as long as the cell inside the prison is above ground (in the analogy, the curious rule of the new freedom says the digging of the tunnel has to proceed horizontally). Furthermore, if the particles are successful in getting past the barrier, they will roam freely with whatever energy they had before their transit to freedom.

A marble resting undisturbed in the bottom of a glass atop a table will stay there forever according to classical physics, but the subtleties of uncertainty and complementarity say there is a possibility that it will spontaneously materialize on the outside of the glass and roll away, without having received any external impetus. One physicist at a Royal Society meeting later that year illustrated the situation facetiously, saying after a talk by Gamow, "Anyone present in this room has a finite chance of leaving it without opening the door, or, of course, without being thrown out through the window." He then reassured the audience, telling them that the probability of such an ejection was negligible. But events of this sort are not quite so rare in the nuclear world.

The possibility of quantum mechanical tunneling was already known in 1928, but quantum mechanics was still a very young subject. Now, by the initial application of quantum mechanics to nuclear physics, Gamow had succeeded in explaining many of radioactivity's puzzling features. A low, narrow barrier would lead to an atomic lifetime lasting only fractions of a second (an intensely radioactive element), while a high and deep one meant stability for millions or even billions of years. Under certain circumstances, not at all uncommon (imagine the prisoner's cell having been built below ground), the nucleus would never decay.

During the weeks after his insight, Gamow worked out the details of his theory. Finding that his calculations fit all the experimental data beautifully, he wrote a paper on the subject and submitted it to the famous German journal *Zeitschrift für Physik.*

Gamow's paper on the subject of radioactivity was a very important one, and he wanted to stay longer in Göttingen to expand the work. Unfortunately, he had by then exhausted his limited funds, and it was time to return to Leningrad. But he decided that on his way home he would stop in Copenhagen for twenty-four hours to see the famous institute and perhaps even meet the legendary Bohr. Arriving at Blegdamsvej, he explained to Bohr's secretary in halting German the purpose of his mission. She said he would have to wait a few days before he could have an appointment with Bohr, but on being told that he had only enough money for one day in the city, she disappeared, coming back quickly with the great man himself. Bohr took Gamow into his office and politely asked him what he had been working on. Gamow's poor German was not an obstacle. The equations spoke for themselves.

When he finished, Bohr, immediately recognizing the importance of Gamow's explanation of alpha particle emission, said to him, "My secretary tells me you have only enough money for one day. If I arrange for you a Carlsberg fellowship at the Royal Danish Academy of Sciences, would you stay here for one year?"

Bohr did more than give Gamow a fellowship, which he of course accepted with joy. Realizing that Rutherford should also hear about the young Russian's explanation of alpha particle emission, he arranged for Gamow to visit Cambridge. Since Sir Ernest was wary of complicated theoretical presentations—he used to say he didn't want to hear any physics that couldn't be explained to a barmaid—Bohr sent along a letter telling Rutherford to listen carefully to the young Russian. As insurance, he also shipped him a note showing a plot comparing the fit of Gamow's model and that of Rutherford's two-tugboat model to Rutherford's own experimental results. Since it was clear which one was a better match, Bohr knew it would make Rutherford sit up.

Rutherford liked the ebullient young Slav, and Gamow liked the distinguished New Zealander. When his Danish fellowship came to an end, Gamow went to Cambridge for a year with a Rockefeller Foundation fellowship that Bohr had helped him obtain. The "Slav giant" must have startled even English eccentrics as, sportingly clad in golfing clothes, he cruised around the countryside on a large used motorcycle he had purchased. But his playfulness was balanced by serious, intense work. Among other things, Gamow conveyed an observation to Rutherford that turned out to be very important for the future of nuclear physics.

If internal alpha particles can leak out of a nucleus through a barrier, Gamow realized it might not be that hard for external alpha particles aimed at a nucleus to break in. One could just as easily tunnel into a jail as tunnel out. Though this is never done in the outside world, it might be a very attractive activity in the nuclear realm. Rutherford, who had been thinking along similar lines but lacked the expertise to calculate how likely such an event might be, called Gamow into his office one day and asked him how much energy a proton would need compared to an alpha particle in order to gain access to a nucleus's interior. Gamow remembered answering quickly, "one sixteenth." According to Gamow's memoirs, this is what happened next:

> "I thought you would have to cover sheets of paper with your damned formula."
>
> "Not in this case," I said.
>
> Rutherford called in John Cockcroft and Ernest Walton, two young Cambridge experimentalists with whom he had discussed the experimental possibilities beforehand.
>
> "Build me a one-million electron-volt accelerator; we will crack the lithium nucleus without any trouble," said Rutherford.
>
> And so they did.

But that didn't happen until the summer of 1932. In the meantime Gamow moved back to Copenhagen for a second sojourn, bringing with him the motorcycle and offering rides to anybody brave enough to mount behind him. He even lent it briefly to Bohr, who wanted to see if he could steer it. It was now 1930.

During this second stay in the Danish capital, Gamow made friends with Max Delbrück, there on his own Rockefeller Foundation fellowship. Forming a congenial pair, the two of them discovered a shared interest in practical jokes as well as a love of physics. The latter became evident as they began to work together, analyzing the emission of gamma rays. Since these are composed of photons, the quanta of electromagnetic radiation, their discharge from the nucleus depends on the distribution of electric charges within the nucleus. Delbrück and Gamow's results were of necessity preliminary because the nucleus's structure was still essentially a mystery. Though they did not know it at the time, an essential ingredient in clearing up that mystery was still missing. It was found in 1932, with Chadwick's discovery of the neutron.

On his second Copenhagen stay, Gamow had the very welcome surprise of seeing again his Russian friend Lev Landau, the youngest member of his earlier three-man Leningrad study group. Now twenty-one, Landau, or Dau as he was known, was already a seasoned physicist, having graduated from college at sixteen. Dau, who became Russia's greatest theoretical physicist and one of the twentieth century's major scientific figures, was never intimidated by anything or anybody, not even by Gamow. As the Dutch physicist Casimir remembered, "Landau's was perhaps the most brilliant and quickest mind I have ever come across." This is high praise from someone who knew well both Heisenberg and Pauli!

Dau and Geo, as Gamow was known, added immensely to Copenhagen's high spirits, re-creating their Leningrad high jinks with Casimir often recruited to be the third musketeer: "We formed a trio that was most amusing to ourselves, though not always appreciated by others" was how Casimir recalled their actions. Outsiders often judged their behavior toward their elders as irreverent. For instance, as more than one visitor saw, it was not unusual to find Landau, tired after a talk, lying down flat on the lecture bench "arguing and gesticulating up at Niels Bohr, who was bending over him earnestly trying to convince him he was wrong." Bohr never took offense, but visitors were sometimes startled by the informality. Others might think this was undignified, but Bohr regarded it as simply part of the Copenhagen spirit of free exchange. No barrier to discussions was the rule.

Celebrating such sparring, Delbrück wrote into the Faust skit a scene in which Bohr/the Lord speaks to Landau, gagged and bound to a chair. Since Landau was well-known for holding forth on his views anywhere, to anybody, at any time, many felt this was the only way to silence him:

THE LORD
Keep quiet, Dau! . . . *Now, in effect,*
The only theory that's correct,

Or to whose lure I can succumb
Is

LANDAU
Um!Um-um!Um-um!Um-um!

THE LORD
Don't interrupt this colloquy!
I'll *do the talking. Dau, you see,*

Landau, bound and gagged, being lectured to by Bohr.

Bohr enjoyed his exchanges with Landau, but saying what came to your mind was a dangerous habit in the Soviet Union of the 1930s, one that eventually caught up with Landau on his return to the motherland. In 1938 he was put in jail, languishing in a cell for a year. He was saved from almost certain death through the courageous intervention of Pyotr Kapitza, the Soviet Union's most prominent experimental physicist. Kapitza went directly to Vyacheslav Molotov, the chairman of the Council of People's Commissars, telling him in no uncertain terms that he would do no military research for the Soviets if Landau was not released.

Gamow's life also took some precipitous turns in his homeland. In the spring of 1931, having been away for almost three years, he returned to the Soviet Union hoping to get his passport quickly renewed and then go to Rome to attend a conference on nuclear structure. But he found his native country less flexible than it had been a few years earlier. Stalinist ideology had clamped down—Russian scientists were encouraged to discover the secrets of capitalist science while not revealing those of their own proletarian science. They were also ordered not to fraternize with Western researchers. Matrix mechanics, declared antimaterialistic, was not to be used, but wave mechanics was acceptable as long as one did not invoke the uncertainty principle.

As Gamow points out, this sort of nonsensical intrusion of politics into science did not have a long-term deleterious effect on physics research in the Soviet Union, but the corresponding dominance in biology of T. D. Lysenko, with his rejection of chromosomal inheritance, effectively stifled genetics investigations there for years. And the debate about political intrusion into scientific research still rages in all corners of the world.

Gamow's 1931 request for a new passport was denied; he was now trapped. Max Delbrück gave the intended Rome conference lecture in his place. One year later, disappointed that Gamow still could not leave Russia, Delbrück put his personage into the skit as well. As described earlier, Gamow's picture appears behind bars while a voice from the rear says,

I cannot go to Blegdamsvej
(Potential barrier too high!)

This was a clear double reference to Gamow's discovery of how alpha particles exit from nuclei and his political problems in trying to leave the Soviet Union. In October 1933 Gamow and his wife serendipitously obtained exit visas to attend the seventh Solvay conference. They never went back.

Heaven and Earth

DURING HIS SECOND Copenhagen stay, Gamow decided it was time to present his thoughts in book form. The result, published in 1931 by the Oxford University Press, appeared under the title *Constitution of Atomic Nuclei and Radioactivity.* Written by a twenty-six-year-old, it was the world's first nuclear physics text. Given his terrible spelling and grammar, the publisher hired a physicist to translate Gamow English into English English; it was hard work, although the editor admitted, "There is an occasional correct sentence."

Gamow was convinced of the truth of his own observations about nuclear decay by alpha particle emission but had serious doubts about explanations of the mechanisms that produced beta rays, as proposed by him or by anybody else. To alert his readers, he had a special rubber stamp of skull and crossbones made, using it to mark passages in the book's galleys that dealt with the topic. The publisher suggested using a tilde instead, remarking that it was a slightly less ominous notation. Gamow reluctantly agreed, answering, "It has never been my intention to scare the readers more than the text will undoubtedly do."

Skull and crossbones.

In a sense the problem physicists faced with beta rays was the reverse of the one encountered with alpha rays. Beta rays' constituents were known with certainty to be electrons and nothing more, while alpha rays' constituents were still somewhat of a mystery. Rutherford had proved more than twenty years earlier that alpha particles are helium nuclei, but what are they made of? Two protons and what else?

On the other hand, though alpha rays' mode of exit from the nucleus was well understood, how and why did beta rays come out of the nucleus and why was energy apparently not conserved when they did? Experiments indicated that either energy was truly not conserved in beta ray emission or something undetected accompanied the rays when they exited from the nucleus. Bohr chose the first option and Pauli the second, but neither point of view seemed satisfactory, prompting Gamow's 1931 display of skull and crossbones.

Many thought Bohr's proposal was nothing more than a return, in a new guise, to his discredited BKS (Bohr-Kramers-Slater) notion, the one Einstein had labeled "an old acquaintance of mine but not an honest fellow." Bohr had maintained in BKS that energy is only conserved on the atomic scale when averaged over many events, but not true case by case. Experiment had proved Bohr's suggestion wrong, and he had accepted defeat graciously. Five years later, he seemed to be making the same argument, except now on the much smaller scale of the atom's nucleus.

Bohr was aware of the criticism, and if beta rays had been the only problem, he probably would not have raised the issue again, but three other mysteries were on his mind. He thought they might somehow all be connected to energy nonconservation.

The first of these was the mechanism that kept stars shining. This had been a major science concern at the end of the nineteenth century, because conventional ways of generating energy could not explain the long solar life necessary for biological life to evolve. The solution had supposedly been found with the discovery of a novel source of energy, radioactivity. However, by 1929 it had become clear that something even more powerful than radioactivity was needed to account for the predicted age of the sun. Bohr now thought that the existence of giant stellar interiors of compacted nuclei in which energy was not conserved might provide the needed stellar luminosity.

Bohr's Copenhagen assistant Oskar Klein discovered the second problem. At the end of 1928, examining closely Dirac's union of relativity and quantum mechanics, Klein noticed how it predicted a curious behavior for electrons striking a barrier. Physicists expected that some electrons would penetrate the barrier and others would be reflected, as Gamow had shown for alpha particles. But Klein noticed that Dirac's theory indicated that more electrons would be reflected than would hit the barrier in the first place. In other words the number of electrons was not conserved. Bohr now asked himself if number nonconservation could possibly be related to energy nonconservation?

The third conundrum, commonly called the wrong-statistics problem, concerned the picture of electrons and protons bound tightly to one another within nuclei. Nitrogen, with atomic number 7 and atomic weight 14, supposedly had a nucleus containing 7 protons and 7 electron-proton pairs, 21 particles in all, an odd number. Experiment showed the total number of particles inside that nucleus had to be even. Something was at fault here as well.

Though Bohr didn't propose the idea of energy nonconservation in a formal lecture until May 1930, he had been testing it with colleagues for well over a year, usually encountering strongly negative reactions, a reflection of how deeply seated the principle of energy conservation was. In February 1929, Pauli was already complaining to Klein about Bohr's being on the "wrong track," attributing part of the blame to Bohr's "helplessness in the discussion of experiments." He also chided Bohr: "Do you intend to mistreat the poor energy law further?" he wrote, and later urged him to "let the stars shine in peace." Writing again to Klein, Pauli said he was making a trip to Berlin to speak to "Fraulein Meitner for the purpose of gathering evidence against the Copenhagen theoretical nonsense."

The issue of energy nonconservation was debated vigorously at the first Copenhagen meeting, held in April 1929. Despite Pauli's opposition, Bohr could not be dissuaded from his views. Over the summer he prepared a first draft of a manuscript on the subject and sent it to Pauli, whose reaction was predictably negative. Trying once again to convince Bohr not to go any further, Pauli wrote to him, "Let this note rest for a good long time."

Not giving up, Bohr sent Dirac a letter in which he proposed a possible link between energy nonconservation at the nuclear level and the apparent nonconservation of electrons that Klein had discovered. He was rebuffed again, as Dirac answered, "My own opinion of this question is that I should prefer to keep rigorous conservation of energy at all costs and would rather even abandon the concept of matter consisting of separate atoms and electrons than give up conservation of energy." As usual, Dirac was very clear on where he stood.

Probably informed by Dirac of the state of affairs, Rutherford sent his Danish friend a message containing an allusion to remarks by another famous Dane, Hamlet's speech to Horatio:

> I have heard that you are on the warpath and wanting to upset the conservation of energy both microscopically and macroscopically. I will wait and see before expressing an opinion but I always feel "there are more things in Heaven and Earth than are dreamt of in our philosophy."

Hamlet's remark turned out to be the wisest. Though nobody could have foreseen it, the solution to Bohr's four problems required the existence of three new kinds of particles (neutrinos, positrons, and neutrons), two entirely new forces in nature (the strong and the weak), major theoretical advances explaining how these forces operate, and the tracing of nuclear reactions in the stars by means unimaginable in 1930.

Bohr had been looking for one common answer to four problems. They instead had four unrelated answers, each one dramatic and each one changing the course of physics.

The Revolutionary Proposals

FOUR BIG PROBLEMS. Some physicists began to have a feeling of foreboding akin to what they had experienced five years earlier as the Bohr model's inadequacies became increasingly evident. Those worries had been swept aside by the triumphs of matrix mechanics, wave mechanics, and the Copenhagen interpretation of quantum mechanics. Now they were slowly returning in another guise.

Physicists started asking themselves if they now had the right theories. The negative energy solutions of Dirac's equation, the beta decay problem, and other mysteries weighed on their minds.

This disquiet was still present at the 1932 Copenhagen meeting, visible in the skit's adaptation of the scene toward the end of Goethe's drama where four gray women appear to Doctor Faust, announcing themselves as Want, Guilt, Care, and Need. In the parody, the four gray women are electron-theory problems. They lead the character called Dirac to declare,

Our theories, gentlemen, have run amuck.
To 1926 we must return;
Our work since then is only fit to burn.

This was an exaggeration. Physicists did not feel their work *fit to burn*, but how to proceed was far from clear.

Toward the end of 1929, Dirac had proposed a solution to the problem embodied by the third of the gray ladies, Negative Energy. It struck many as outrageous. The question had been, Why don't electrons drop from positive energy states to negative energy ones as surely as a ball rolls down a hill? Dirac's solution was to assume that every single one of the infinite number of negative energy states his equation predicted was occupied. Since according to the Pauli principle no two electrons can be in the same state, there was no chance a positive energy electron could drop into a negative energy state; these places were already taken by preexisting negative energy electrons.

How could we possibly not notice an infinite number of electrons with negative energy? Dirac's answer was a throwback to the Pythagorean scheme that inspired Kepler. Why doesn't one hear the harmony of the spheres in the Pythagorean cosmos, the ensemble of music emitted by the planets circling the central fire? Pythagoras said we are unaware of it because that music is the constant background of our lives. As Shakespeare wrote, giving voice to a notion still lingering two thousand years after Pythagoras's death,

> There's not the smallest orb which thou behold'st
> But in his motion like an angel sings,
> Still quiring to the young-eyed cherubins;
> Such harmony is in immortal souls;
> But, whilst this muddy vesture of decay
> Doth grossly close it in, we cannot hear it.

Likewise Dirac said we *cannot hear* the electrical noise of the forever-present infinite number of negative energy electrons because it has always been with us.

Dirac's next step was an option not available in the Greek world of eternal and unchanging heavens. Nevertheless Pythagoras would probably have agreed that if one of the spheres were suddenly removed from the sky, we would perceive its absence from the ensemble as the presence of a melody. In other words, either an addition or a deletion to a constant background is heard as music.

In this vein, Dirac suggested that the *absence* of a negative-energy, negative-electrical charge electron should be interpreted as the *presence* of a positive-energy, positive-electrical-charge particle; conjecturing this might provide an explanation for the proton's existence. He said perhaps there is only one kind of particle; we call its positive energy manifestation an electron while a proton is simply the absence of its negative energy counterpart.

Most physicists could not accept such a notion. Dirac's infinite sea of negative energy electrons with absences or, as he called them, *holes*, seemed too fanciful, even granting that Dirac was a genius and mathematical consistency had been his guide. Pauli, who found the whole notion unattractive, was particularly annoyed by Dirac's appeal to Pauli's own exclusion principle as a means of doing away with the negative energy states. On the other hand, there was no question that the Dirac equation seemed to be correct. What was one to do?

Nor was an accepted solution to how beta ray emission occurred anywhere in sight. In July 1929 Pauli, increasingly irritated by Bohr's talk about energy nonconservation, wrote him from Zurich that he had just heard a lecture by Fraulein Meitner that convinced him more than ever that there were real problems in beta decay: "We really do not know what happens here. Neither do you, you are able only to indicate reasons for why we understand nothing!"

But Bohr was persistent. Just how persistent he could be was certainly well-known to Heisenberg and Pauli. As an aside, there is an interesting exchange on this subject between Pauli and his Zurich friend Carl Gustav Jung. During the course of a twenty-five-year correspondence, Pauli sent Jung summaries of over a thousand of his dreams. Trying to explain to Jung the importance of Bohr's appearance in one of them, Pauli tells Jung about Bohr's powers: "In real life he does actually have the ability to overcome people's resistance." In this case Bohr did not overcome Pauli's resistance, but did pressure him to provide a better answer to the suggestion of energy nonconservation than *we understand nothing*.

On December 4, 1930, Pauli made his counterproposal in an open letter to a group of radioactivity experts who were meeting in Tübingen, a well-known German university town. With a keen sense of history in the making, Meitner, who attended the meeting, kept a copy of the letter. It reads in part,

> Dear Radioactive Ladies and Gentlemen,
>
> As the bearer of these lines, for whom I pray the favor of a hearing, will explain to you in more detail, I have, in connection with the "wrong" statistics of the N and Li 6 nuclei as well as the continuous beta spectrum, hit upon a desperate remedy for rescuing the "alternation law" of statistics and the energy law. This is the possibility that

there might exist in the nuclei electrically neutral particles, which I shall call neutrons, which have spin ½, obey the exclusion principle and moreover differ from light quanta in not traveling with the speed of light. The continuous beta spectrum would then be understandable on the assumption that in beta decay, along with the electron a neutron is emitted as well, in such a way that the sum of the energies of electron and neutron is constant....

I admit that my remedy may perhaps appear unlikely from the start since one probably would long ago have seen the neutrons if they existed. But "nothing ventured, nothing won" and the gravity of the situation with regard to the continuous beta spectrum is illuminated by a pronouncement of my respected predecessor in office, Herr Debye, who recently said to me in Brussels "Oh, it's best not to think about it at all—like the new taxes." One ought therefore to discuss seriously every avenue of rescue. So, dear radioactive folk, put it to the test and judge. Unfortunately I cannot appear personally in Tübingen, since on account of a dance which takes place in Zurich on the night of the 6 to 7 December, I cannot get away from here.

Your most humble servant, Wolfgang Pauli

I want to immediately clear up one source of confusion. Pauli called his new particle *neutron*, but that name was soon also used for a different particle, the heavy neutral companion of the proton within the nucleus. Bewilderment ensued, and an alternate suggestion for Pauli's very light particle was proposed by Fermi and quickly adopted. Changing big to small in Italian involves switching from *-one* to *-ino*, so Pauli's *neutrone* became a *neutrino*. *Neutrino* is the name we still know it by, and the name I will henceforth use in this book.

In his letter Pauli emphasized that every time an electron came out of the nucleus, a neutrino came out as well, carrying away the supposedly missing energy. In his view the neutrino was so hard to observe that thirty years of beta decay experiments might simply have missed detecting it.

Pauli also suggested that neutrinos could provide a solution to the wrong-statistics problem, the third of Bohr's four troubles. Experiment showed that the nitrogen nucleus has an even number of particles. Fourteen protons and seven electrons made twenty-one particles, an odd number. Now add seven neutrinos and the answer changes from twenty-one to twenty-eight, from an odd to an even number.

How radical was Pauli's suggestion? In the beginning of the twenty-first century, we have expanded the scope of our thinking to the point that we now envision matter as made of, inter alia, leptons, which include electrons, muons, neutrinos, and, furthermore, six different kinds of quarks, each in three colors; the notion of one more type of particle seems straightforward. But in 1930 all matter was presumed to be composed simply of electrons and protons. Pauli had said, "I was a classicist and not a revolutionary," but proposing a new particle in 1930 was perhaps a more revolutionary step than suggesting energy is not conserved within the nucleus.

It is possible that he had loosened his normally rigorous standards and speculated in a way he normally would not have because of the tumult in his private life. His divorce, after less than a year of marriage, had become official on November 26, 1930. His proposal for solving the beta decay energy-conservation problem appeared a week later. Reminiscing about that time in a letter to Delbrück written shortly before his death, Pauli described the neutrino as "that foolishly behaving foolish child of my life's 1930–31 crisis."

That crisis in Pauli's life was very real. The years of transition from youthful prodigy to established professor had been very hard for him. His mother's suicide, his unhappy marriage and quick divorce, and his leaving the Catholic Church all conspired to form a combination that led to excessive drinking, smoking, eating, and a variety of neuroses. Perhaps physics provided him with an island of stability during that time, because Pauli maintained both his interest and his remarkable productivity in physics throughout his troubles.

The smoking and drinking, however, caused Pauli some problems, never more so than during a 1931 three-month visit he made to the

United States, then in the grips of Prohibition. In a letter to a Zurich friend, he lamented the "small bourgeois, Philistine" aspect of America where a dinner would conclude with prayers rather than "coffee and cigars—not to speak at all of alcohol." On a stay in the vicinity of Canada, at the University of Michigan, he crossed the border to make up for the earlier abstinence on the trip. The resulting gratification had some problems of its own. Though the official story was that Pauli had slipped stepping out of a boat, the truth was that he had tripped on a step while he was, as he put it, "slightly tipsy." The resulting broken bone in his shoulder required his right arm to be immobilized in a cast set at forty-five degrees to his body. Nevertheless, as he spoke facing his audience, with a friend writing on the board the accompanying equations as directed, Pauli's lectures had never been better. He felt that the worst part of the arrangement was that the cast kept his arm at an angle such that it mimicked the Nazi salute.

The Three Young Geniuses Each Write a Book

HEISENBERG WAS HAVING his own difficulties, feeling sad as he turned thirty in December 1931. There were piano playing, exchanges with colleagues and students, skiing, hiking, and even horseback riding, but he felt something central to his life was missing. In large part, it seemed to be the loss of the Neupfadfinder camaraderie. That pure, uncomplicated youth movement had sustained him in his earlier years, providing an alternative to the intense world of physics research. He was now part of the Neupfadfinder's *Altmannen*, the older men or alumni, but they were relatively inactive. Furthermore, the Nazi Party was espousing a vision perilously close to what his brotherhood had once strived for, a new, untainted German Reich guided by Teutonic morals. He could only watch as his utopian vision was turned into a nightmare of totalitarianism, fear, and repression.

Matters were certainly not helped by Germany's economy, which had collapsed once again, brought down by the worldwide depression that followed the Wall Street market crash. As discontent mounted, the unemployed, the underemployed, the disgruntled, and the beleaguered began demanding answers. Hitler provided them with a response: it was the fault of Jews, Communists, war profiteers, the Treaty of Versailles, and traitors to the fatherland. His rise paralleled the economy's decline. By September 1930 the Nazi Party, once a small splinter group, was the second largest in Germany.

With the pall from the world news becoming increasingly hard to shake, Heisenberg's optimism about the future of physics also seemed to be crumbling and, for once, so was his self-confidence. His initial elation at achieving so much and being recognized had faded. By mid-1931 he was unhappy enough to write Bohr, "I have given up concerning myself with fundamental questions, which are too difficult for me." Nor was he convinced a change would come soon. Heisenberg now thought progress in physics would require a thorough understanding of the atomic nucleus, perhaps as big a step as quantum mechanics itself had been, and saw no way that he or anybody else could take that step.

Dirac, the third of the three young geniuses in the Copenhagen front row, appeared unperturbed. Continuing his solitary walks and his life as a don in a Cambridge college, he seemed almost ageless, guided as always in his thinking by criteria of mathematical elegance, logical consistency, and style. But he was also having problems, though they were of an intellectual, not an emotional, sort.

Dirac had hoped those so-called holes, absences from an infinite sea of unobserved negative energy and negative charge, might be protons, but in 1930 several authors independently showed this could not be. In early 1931 Dirac admitted that the holes had to correspond to particles having a mass identically equal to an electron's, but with an opposite electrical charge. They would be called antimatter particles. If his theory was right, antielectrons had to exist and act in such a way that if one of them encountered an electron, they would both disappear, leaving behind only a flash of radiation. Like Pauli, he too was now predicting a new kind of particle, one even more bizarre than the neutrino.

Pauli was predicting the existence of neutrinos, Dirac was saying antielectrons exist, and Heisenberg was maintaining that nothing was good enough. Reeling, the physics world thought, What comes next?

As the three geniuses moved forward once again into the unknown, they also acted to confirm to the world of physics how far it had come in the past five years. Whatever concerns they had for the future of their discipline, quantum mechanics stood on solid ground, and it was time for everybody to know that. Heisenberg, Dirac, and Pauli each sat down to write in book form his own version of the theory. It would be the harbor from which each would set sail again.

The three books demonstrate three different styles, three ways of looking at the same problems, each in its own way a vivid self-portrait of its author as well as a commentary on the state of the craft. As one might expect given the rapid pace of Heisenberg's thinking and actions, his book was the first to appear. Based on lectures given at the University of Chicago in the spring of 1929, it is clear and short, announcing at the start that its purpose is to explain the Copenhagen *Geist* (spirit) of quantum theory.

Pauli's book, appearing in 1933, is the last and most thorough of the three. He had come to the attention of the physics world by his magisterial all-encompassing review of relativity theory, the one Einstein had praised so lavishly. Pauli had shown then his ability to absorb vast amounts of material, extract the cogent points, and present the whole subject panoramically. Pauli, the arbiter of quantum mechanics, was perhaps the only physicist whose knowledge ranged from the most arcane mathematical developments to the intimate details of experiment. Knowing the ins and outs of all quantum mechanics arguments, he conversed equally easily with Dirac and with Meitner, with theorists and experimentalists.

Pauli had already provided in 1926 a three-hundred-page overview of quantum theory for the *Handbuch der Physik*, an encyclopedic reference book of those times, in which long detailed articles by experts appeared. He decided in the early 1930s to write what he called the "New

Testament." It was intended as a companion to this earlier publication, which would then be known as the "Old Testament." Finished in late 1932 and published in the next year, Pauli described it to Hendrik Casimir, his assistant at the time, as "not quite as good as my first *Handbuch* article, but in any case better than any other presentation of quantum mechanics." Many physicists agreed with that latter judgment but thought his New Testament at least as good as his old one.

Heisenberg's book was a good way to get an introduction to quantum mechanics, Pauli's was the best for learning the details of the subject, but only one of the three texts is still widely read today, Dirac's classic 1930 *The Principles of Quantum Mechanics.* There are better places and easier ways to learn the subject, but this book is, like most of this strange man's creations, stylistically perfect. This very perfection occasionally caused dismay in Cambridge when his lectures on quantum mechanics were found to consist of nothing more than him reading from his book. When asked why he was repeating its contents verbatim, he replied that he had given the matter considerable thought and what he had written represented the best way to express his conclusions.

For one who knows quantum mechanics, the book is still a joy to read. (In the interests of full disclosure, I should add that it is the only one of the three I have read from cover to cover.) In its preface, contrasting the quantum view with the one that originated with Newton, Dirac says,

> It has become increasingly evident in recent times, however, that nature works on a different plan. Her fundamental laws do not govern the world as it appears in our mental picture in any very direct way, but instead they control a substratum of which we cannot form a mental picture without introducing irrelevancies.

He then states some of the problems faced in understanding that "different plan," the one nature works on, the one "of which we cannot form a mental picture."

> The new theories, if one looks apart from their mathematical setting, are built up from physical concepts which cannot be explained in terms of things previously known to the student, which cannot even be explained adequately in words at all. Like the fundamental concepts (e.g. proximity, identity) which every one must learn on his arrival into the world, the newer concepts of physics can be mastered only by long familiarity with their properties and uses.

The writing is beautiful and the logic always impeccable. For a physicist, this is a true work of art. No other presentation explains quite so succinctly why quantum mechanics is difficult, nor does another text make this theory's elegance quite so obvious.

The different attitudes of these three books toward Bohr are also telling. Heisenberg, perhaps feeling guilty about the acclaim coming to him for the uncertainty principle, does not even refer in his bibliography to his own paper on the subject, instead crediting Bohr lavishly, almost saying uncertainty is a by-product of complementarity. As expected, Pauli discusses in detail the features that pertain to uncertainty and those that relate to complementarity, assigning credit where credit is due. By contrast, Dirac has a section on "Heisenberg's principle of uncertainty," but does not even mention complementarity, nor does he refer to Bohr other than as a contributor to the old quantum theory. The Dane's contributions simply do not fit into his logical presentation. Dirac had developed his own understanding of complementarity; in his own words, he felt that Bohr's version "doesn't provide you with any equations you didn't have before."

By the time physicists of my generation began studying the subject, Bohr's contribution to quantum mechanics' development was largely ignored even though we knew the subject's form went by the name of the Copenhagen interpretation. Quantum mechanics was regarded as hard enough and as having so many technical and conceptual spin-offs that we felt we couldn't afford to dawdle by exploring its historical fine points. Our summary view consisted of an abridged early history with Planck, Einstein, and Bohr figuring prominently. Then came the revolution, founded on Pauli's exclusion principle and Schrödinger's equation (with Heisenberg finding the same results in an inscrutable way whose details we didn't examine); Heisenberg's uncertainty principle and Dirac's equation followed.

We heard that there had been a decades-long debate between Bohr and Einstein about what it all meant, but since they hadn't settled it, what chance did we have to understand what they had argued about? We thought that the few members of our generation who did worry deeply about quantum mechanics were to be admired but not emulated. We were going to solve problems that had numbers as answers, make predictions, explain experiments.

As I taught quantum mechanics and read about the early days of its formulation, it came as somewhat of a revelation to find what a key figure Bohr was, how he led, prodded, stimulated, challenged, and united the younger theoretical physicists, how he created for them an atmosphere where the very best in them would be drawn out. In the process I have also come to understand why they loved him and why so many referred to him as the dominating influence in their intellectual lives. In short, I have come to appreciate why he was, facetiously and perhaps not facetiously, the Lord.

Chapter 12

The New Generation Comes of Age

STUDENT
Profoundly learned I would grow,
What heaven contains would comprehend,
O'er earth's wide realms my gaze extend,
Nature and science I desire to know.

MEPHISTOPHELES
You are upon the proper track, I find;
Take heed, let nothing dissipate your mind.

Goethe, *Faust, Part I,* 1543–48

The Apprenticeship

AS THE YOUNG MEN and women finishing their physics studies in the late 1920s began looking for mentors to help them enter the field, they also came to know each other, to work and play together, to weave informal networks, connections that might be just as important as the guidance they received from elders. Göttingen, Berlin, Munich, Leiden, Copenhagen, Leipzig, Zurich, and Cambridge were a carousel for them, and they hopped on and off as their assistantships and fellowships allowed, only to leave it in their late twenties when the time came to find a more permanent settling place.

Looking for places to perfect their skills, the young were acutely aware of the changes taking place. Sommerfeld, now sixty years old, was still active, but not the force he had once been. Born was withdrawing from the fray, and Ehrenfest was increasingly subject to bouts of depression. They couldn't turn to Schrödinger and Einstein: neither of them took students, nor did they fully accept the triumphant tenets of quantum mechanics' Copenhagen interpretation.

By the beginning of the 1930s, Bohr was the only member of the older generation still in the thick of the chase, the only one to be fully conversant and engaged with the new developments in theoretical physics. This enhanced his symbolic importance as a father figure, his very real influence as a link to the past, and the power of his vision for the future. Copenhagen, more prominent than it had ever been, became the mecca of theoretical physics.

But it wasn't necessarily the right place for all young theoretical physicists, as J. Robert Oppenheimer learned. After graduation from Harvard College in 1925, he had gone to Cambridge. Not happy there, he moved on to Leiden and then, at Ehrenfest's suggestion, to Göttingen, a stay that proved very productive for him. In the fall of 1928, after a year in the United States, he was back once again in Leiden, thinking Copenhagen should be his next stop, but Ehrenfest intervened. Feeling that Oppenheimer needed rigor in his thinking more than long discussions with Bohr, he told him in no uncertain terms to go instead to Pauli in Zurich. Ehrenfest then wrote Pauli telling him that the young man was very talented but needed looking after by the one and only *Geissel Gottes* (Scourge of God).

The prescription was effective. A few months after his arrival in Zurich, Pauli reported back to Ehrenfest that Oppenheimer was doing well, working personally with him. However, the young American had, in Pauli's words, "a very bad quality: He approaches me with a fairly absolute authority cult and considers all I say . . . as last and definitive truth." Pauli then described how he had tried to break Oppenheimer of this conviction and closed the letter by telling Ehrenfest this:

> One thing, however, I hope to have achieved soon: that Oppenheimer, at least in relation to me, adopts my manners! This is absolutely necessary if I should not consider myself a rascal. For formal politeness, in my view, is the great heresy which needs to be ruthlessly eradicated from human relations—this is an unshakeable dogma. —signed— The Scourge of God

In later years Oppenheimer was in fact known for his biting sarcasm, perhaps not at Pauli's level, but significant enough to be noted by many. However, it was not always viewed as kindly, because Oppenheimer, more than Pauli, often came to be seen as arrogant. On the other hand much of this flowed from Oppenheimer's later deep engagement in the worlds of politics and of the military, domains that Pauli never embraced to the slightest degree.

Oppenheimer learned in Zurich and in his earlier stays in Göttingen and Leiden the importance of a having a powerful group of theoretical physicists working together, teaching and learning from one another. On his return to the United States, he established two great schools of theoretical physics in California, one at the university in Berkeley and the other at the California Institute of Technology in Pasadena. These were two of several centers that emerged in the 1930s to spread the doctrine of quantum mechanics' Copenhagen interpretation. In the process, they helped to make the United States a world leader in physics research.

Individuals going back to their homelands from Europe preached the new quantum mechanics gospel. Sometimes the sermon was delivered by immigrant European physicists; occasionally the advance of the new physics was simply a matter of talent spontaneously arising from supposedly barren soil, just as Gamow and his friends had educated one another in Leningrad.

The most influential of the new research hubs of the 1930s was in Rome, the Eternal City. While Heisenberg, Pauli, and Dirac were establishing themselves in Leipzig, Zurich, and Cambridge, a fourth member of their generation was also becoming known as a leader and educator. In Rome, physicists called Enrico Fermi *Il Papa,* the Pope, because he was regarded as infallible. A little younger than Pauli and a little older than Heisenberg, he did not grow up in an academic family or have the benefit of early teachers like Sommerfeld and Born. Like Dirac, he was essentially self-taught, perhaps even more so since Italy was a physics backwater while he was first pursuing his career.

Fermi won early fellowships for brief stays in Göttingen and Leiden, the latter particularly important because of Ehrenfest's strong encouragement, but at age twenty-four he was back in Italy. There he developed his own problem-oriented, rather than principle-oriented, style, first rising to world prominence in physics with a 1926 paper in which he applied the ideas of quantum theory, including the Pauli principle, to statistical mechanics. In 1927, having been appointed to fill Italy's first chair of theoretical physics, he began attracting disciples from his own country and elsewhere. They were eager to gain knowledge from the new master and to have the pleasure of living in Rome and seeing its sights, still relatively unchanged from those that Goethe had admired 150 years earlier.

By 1930, with Munich and Göttingen in decline, the preferred path for a young theoretical physicist was to spend time working with Dirac, Heisenberg, Pauli, and Fermi—and of course Bohr. They were the senior figures to learn from. That learning might not always be direct; the young knew that Dirac, for instance, was not very talkative and did not believe in collaborating with others, but the prestige of each of these individuals was a guarantee that the surrounding atmosphere would be stimulating.

The personalities of the new gathering places were different from those of the old, in part a reflection of the changing times and in part because these masters, with the exception of Bohr, were only a few years older than their students and, except for Fermi, still single. Admittedly Pauli had suffered through a brief and disastrous marriage, but that was over. The reach of Copenhagen could be seen in the new informality. Professor and student were more likely to have similar perceptions, to be sharing the same angst about their futures, to socialize as quasi-equals. Young physicists would be drinking beer with their mentors on Sunday evenings rather than having Frau Professor pour tea for them in a proper bourgeois home. One cannot imagine an assistant of Born's or Sommerfeld's playing table tennis with either of them, as Heisenberg's did with him, or filling the kind of role Pauli's first assistant in Zurich later recalled: "One of my tasks, not agreed upon beforehand, was to watch out that Pauli should limit his consumption of ice cream at Sprungli's Konditorei at the Paradeplatz where we often went in the afternoon."

Rudolf, later Sir Rudolf, Peierls was a perceptive student of the four young physics geniuses. Born in Berlin in 1907, he enrolled at its university in 1925 but soon decided he would be better off in Munich. When Sommerfeld left to spend a year in the United States in 1928, Peierls transferred to Leipzig to work with Heisenberg. From there he moved to Zurich as Pauli's assistant and after that obtained a one-year Rockefeller Foundation fellowship, which he divided between Rome and Cambridge, Fermi and Dirac. In a memoir, Peierls provided snapshots of the four.

His thoughts about Pauli and Dirac, similar to observations we have already encountered, underline why both were hard to emulate. Pauli "would insist on getting to the bottom of the problem, and would not stand for any sloppy or half-baked argument." Peierls characterized Dirac's way of speaking as being unlike anyone else's: "You often realized that he followed the question in a straight line, while others, by habit, turned a corner."

But Peierls also observed why Heisenberg, seemingly more conventional than Pauli or Dirac, was also a challenging role model. Describing Heisenberg's approach, he wrote,

> When he was faced with a problem, he would almost always intuitively know what the answer would be and then look for a mathematical method likely to give him that answer. This is a very good approach for someone with as powerful an intuition as Heisenberg, but rather risky for others to imitate.

Nor, according to Peierls, was Fermi's style any easier to copy:

> His methods were always simple, and he did not like complicated techniques. When a problem became complicated, he lost interest. But it must be explained that in Fermi's hands, problems that had been terrifyingly complicated for others often became very simple.

What was the message Peierls came away with? At the risk of oversimplifying, the recipe is (1) never be shoddy or slapdash (Pauli); (2) be logically consistent (Dirac); (3) try to guess the right answer (Heisenberg); and (4) keep it simple (Fermi).

Not many physicists can achieve the right mixture of these ingredients, but Peierls came close. He had a magnificent research career, most of it in England, first at Birmingham and then at Oxford. More than that, he was also a great teacher and mentor. This was particularly evident to me on a spring evening in 1987. I had at the time the good fortune to be spending a semester in Oxford (physicists' wanderings are curtailed but not terminated after their apprenticeship). That day I had been witness to a moving ceremony. To celebrate the eightieth birthday of Prof, as Peierls was known, his former students gathered for a celebration, accompanied by their own students and their students' students. The whole Peierls physics family had assembled. It was a glorious event, a celebration of the man, of his lifetime's work, and of the importance of community in what sometimes is a solitary profession, but need not always be.

The many physicists who had come together that day were reminded of being all part of that great arc of scientific progress stretching back to the beginnings of quantum theory and beyond. As the evening drew to a close and we filed out onto the dark streets of Oxford, we felt embraced by that community.

As the ceremony at ancient Oxford reminded me, the apprenticeship system has not changed very much since the 1930s. Beginning my own career thirty years after the 1932 Copenhagen meeting, I spent a year in Rome, a year in Cambridge, Massachusetts, two years in Geneva, Switzerland, at CERN (the European Organization for Nuclear Research), and two years in Berkeley, California. It was then time for me to step off my carousel, but the people I met in those years are still my friends, my community. And I can only hope my own students are as fortunate.

Copenhagen 1932

PEIERLS WAS NOT the only one to take the theoretical physics Grand Tour. Many years later, by now a Nobel laureate, Felix Bloch recollected his arrival at the Blegdamsvej institute; it was the fall of 1931. Quickly ensconced in two small rooms under the institute's roof, he realized these were the rooms first occupied by Heisenberg, the rooms where Bohr and the young German had argued late into the night about the meaning of quantum mechanics. Those discussions had taken place only four years earlier, but so much had happened in physics during the intervening years that it seemed like a different world. Heisenberg and Pauli, then ambitious youngsters, were now established professors. The march of time was accentuated by Bloch's arriving in Copenhagen having already been Heisenberg's doctoral student in Leipzig and then Pauli's assistant in Zurich. Bloch was now twenty-six, the same age as Heisenberg during his famous discussions with Bohr.

Fresh from working with the two young masters, Bloch remembered meeting Bohr a few days after his arrival. He was descending the spiral staircase that led from his rooms to the institute's library when the great man greeted him:

> "Do you also belong to that gang?" It was fairly obvious that he referred to George Gamow, Max Delbrück and their associates, who had just cleared the field, and whose practical jokes were rapidly becoming a legend all over Europe. Nevertheless, to be quite sure, I asked him whether he meant those people who took nothing seriously and did not respect anybody. Then Bohr smiled, and said in the most wonderful plural of majesty: "Oh, but we do not take their lack of respect seriously either."

Bloch's awe was quickly dissipated by Bohr's obvious warmth and humor. In a few days he found himself slipping comfortably into the Copenhagen spirit of work and play.

It's easier to describe Bloch's way of playing than his way of working, for what does a theoretical physicist do? There is no single answer, just as there is no one answer for what writers or painters do. Research is such a personal matter and reflects so much the character of an individual. Experimentalists depend on equipment, but there is no real equipment in theoretical physics. Physicists jokingly say that all that is needed is paper and pencil, and sometimes even that isn't necessary. Many of the best ideas in theoretical physics were had while hiking, lying on a beach, or listening to a concert—while taking a walk on the barren shores of Helgoland or hiding away with a mistress in Arosa. In theoretical physics there are no services provided to clients and no fixed schedule. While productivity is expected, there are few real ways to measure it. Some of its practitioners—Dirac comes to mind—always prefer to work quietly on their own, and others, like Bohr, feel the need for one or more individuals to be present to provide a context for their musings.

A typical theoretical physicist's day might begin by reading some recent articles in order to keep up to date with new developments. This activity often flows smoothly into meetings with co-workers, students, and visitors. Teaching, talking, and listening are usually a big part of the daily routine, taking place in many venues: lecture halls, informal discussions, casual encounters, and seminars. Because of this, one common feature of almost all theoretical physicists' offices is a blackboard. The rooms themselves may be neat or messy, with or without books, but that blackboard is always there, usually filled with the hieroglyphics of strange equations in various handwritings.

The bonds formed in these exchanges can be very close, sometimes extraordinarily supportive and other times very tense, often oscillating quite rapidly. In the example of Bohr and Heisenberg's relationship, one can find Heisenberg "breaking out in tears because I couldn't stand the pressure from Bohr" and not too long afterward Bohr remembering, "Rarely have I felt myself in more sincere harmony with any other human being."

Sometimes a theoretical physicist's task is quite specific. When an experimentalist arrives in your office and asks you to calculate what a theory predicts for a measurement he or she has just made, the secret or not so secret hope is that the result you obtain will differ from the measured one, causing other theorists to take notice since it may mean something new is on the horizon.

One part of the life is unescapable. Just as a painter must eventually put paint on the canvas, a theoretical physicist has to calculate. The necessary mathematical tools have to be sharpened or learned, often not an easy task, and then applied to the problem in question. The length and difficulty of the calculation varies over an incredible range. If the work is done with collaborators, the task may be subdivided, though any cautious theorist will always make sure all calculations are checked. It is also not a bad idea for everybody to do the same calculation independently.

Reflecting this great diversity, the time from first having an idea to writing a paper can vary from hours to years, though I confess that I do not remember many examples when the time span was hours. Even ideas that seem clearly right to an author are usually mulled over for a few days before their final form is set. Then the paper is submitted to a journal, and there it is refereed, usually anonymously. If it is challenged and sent back, the author may recognize the referee's wisdom or irately resubmit it, calling the referee a fool.

These are some of the ways a theoretical physicist's days pass, though I should not forget attendance at conferences, where one meets one's friends and colleagues, learns what others have been doing, what they plan to do, and what they may only dream of doing. Most theoretical physicists would say their pursuit is a passion, a calling, a joy, not work. It can also be frustrating and disappointing, but few physicists would trade their profession for another one.

A theoretical physicist's style of playing, by contrast, is easier to describe because it diverges less from the amusements of others. But, as in other highly charged professions, physicists' humor may hover on the brink of silliness. Hendrik Casimir, the student Ehrenfest brought to the first Copenhagen meeting, interpreted physicists' occasionally juvenile behavior as a response to stress and the need for mental intensity: "Yet the concentration cannot be maintained over very long periods of time and so one sought refuge in these childish pranks that brought relaxation without interfering with the real work."

Casimir remembers often going to the movies in Copenhagen with Gamow and Landau, usually to see and laugh at action thrillers. Sometimes Bohr would come with them and try to analyze what he had watched. Casimir rendered his impressions of those trips in doggerel verse in a speech he delivered in 1935 for the celebration of Bohr's fiftieth birthday:

> In a Western picture, where guns often bark,
> But it's always the hero who first hits the mark.
> At the end of the movie Niels Bohr, deeply moved
> Set out to explain what the plot really proved
>
> . . . .
>
> The scoundrel must make a momentous decision
> And that interferes with his speed and precision
> But for the defendant there's no such distraction,

Not a shadow of doubt can retard his reaction.

. . . .

This tale has a moral, but we knew it before
It's foolish to question the wisdom of Bohr.

The humor in the verses above and in the Copenhagen skits may be sophomoric, but it opens a window on the feelings and thoughts of young physicists and reminds us that science, even in most abstract forms, is formulated by humans with the same anxieties and frailties we all have.

The Copenhagen performances have another significance. They mark the beginning of a streak of playfulness that has carried into the present, a seeming frivolity in choice of names if nothing else. For example, in the past two decades elementary particle physicists have explained that the visible portion of the universe is chiefly composed of flavored quarks bound in triplets by colored gluons while the unseen and far more massive part is made up of dark matter and WIMPs (weakly interacting massive particles). They hope that the next generation of particle accelerators, multibillion-dollar enterprises, will discover flavored squarks and colored gluinos, the particles' supersymmetric partners.

This type of language can be traced to its mischievous origins back in the Copenhagen of the early 1930s. Some find in the humor that generates these names a lack of seriousness. I see it as a suggestion to be daring, to attack the big problems, to not shrink in front of obstacles, to retain one's childlike curiosity. How else could one have the courage to address issues as deep as the beginning of the universe or the ultimate constituents of matter? This was the secret behind the laughter in Copenhagen.

There were further Copenhagen parodies in the years after 1932, and physicists roared again at the antics of Bohr and Pauli impersonators. In 1937 Bohr took a six-month trip around the world that forced postponement of the meeting from April to September. That year the skit, the one I mentioned my uncle attending, was an adaptation of *Around the World in Eighty Days*. Rather than being the Lord, Bohr was cast as the hero of Jules Verne's classic, Phileas Fogg, except his parody surname was Foggy, appropriate given his soft, almost inaudible voice. In his travels, this globetrotter encounters strange animals—a rhinoceros that speaks like thick-skinned Pauli and a gazelle resembling the ever-swift Heisenberg. That performance was also a big success, but the general agreement was that the 1932 adaptation of *Faust* had been the best skit ever put on at Blegdamsvej.

The "Blegdamsvej Faust"

As usual, the 1932 Copenhagen meeting had no agenda. The day before it was scheduled to start, Bohr simply gathered a few close friends, perhaps only Ehrenfest, Heisenberg, and Kramers. They made a short list of subjects they wanted to have discussed, but the tenor of the gathering was that of a free-ranging debate, with time and opportunity to explore any topic that seemed worthwhile. It was simply colleagues delving into physics.

About a dozen theoretical physicists were in residence at the institute at any given time; they formed the core of the attendees and of the actors in the skit that would inevitably be performed. For the rest, old Copenhageners were invited back as well as a few extras who had something interesting to contribute. Preparing for the 1932 meeting, Bohr knew Pauli would be missed and was sad that Gamow and Landau were trapped in the Soviet Union. But he realized that most theoretical physicists would be thrilled speaking to an audience that included Heisenberg, Dirac, Ehrenfest, and himself in the front row. They would also be reassured to see Meitner there, letting them know about both the reliability of recently published experimental results and what they might hope to learn from experiments in progress.

Bohr once again alerted his friends to feel free to bring along a bright young student. At the first spring meeting, in 1929, it had been Ehrenfest taking the long train trip from Leiden with Hendrik Casimir. It was now Heisenberg's turn to be a mentor. He came to Copenhagen with twenty-year-old Carl Friedrich von Weizsäcker, already well on his way to becoming Heisenberg's closest friend. Weizsäcker's father was a diplomat, and his brother would become a post–World War II president of Germany. The aristocratic young Carl Friedrich, with both a philosophical bent and an experienced view of human affairs, was Heisenberg's guide to the political world. As Heisenberg, still quite unaware of rising movements outside of physics, wrote his mother in 1934, "Only the friendship with Carl Friedrich, who struggles in his own serious way with the world around us, leaves open to me a small entry into that otherwise foreign territory."

In March 1932 Weizsäcker went skiing with the thirty-year-old Heisenberg in the Alps. They talked about their upcoming trip to Copenhagen. Heisenberg, perhaps feeling old, remembered nostalgically how he had gone skiing with Sommerfeld in 1922 and later that year had been taken by him to hear Bohr in Göttingen. He had been only twenty then, the same age as Weizsäcker was now. Then he had been at the very beginning of his career. Now it was his turn to lead the young.

This was Weizsäcker's first Copenhagen meeting. From 1932 on, he went every year until World War II brought the meetings to an end, but this first year was particularly exciting for him. He saw physics being created in front of his eyes, not recorded in a book. One can only imagine how thrilling it must have been to have just turned twenty and to hear directly from Dirac what should or should not be done about antielectrons or to listen to Bohr and Ehrenfest arguing about a fine point in quantum mechanics' interpretation. It was also exhilarating to see physicists only a few years older than he participating fully in the struggle to make sense of the submicroscopic world.

Weizsäcker thought there were so many new puzzles, so many chances to do something important and to contribute to this great enterprise he was still learning about. Was energy conserved or was it not in nuclear decays? How were they going to resolve the curious behavior of nitrogen nuclei? What strange force held nuclei together given that electric repulsion was known to drive its constituents apart? What would their discoveries mean for the world? Though young, Weizsäcker was not naïve and could not help pondering, Would they be a force for good or might there be unintended consequences? How would they know what the future was going to bring?

But for the moment Weizsäcker was still mastering the basic laws of quantum physics. Though not able to contribute to the discussions, he was adept at turning a phrase and was very familiar with the text of *Faust*, so he could at least help with the skit. The script needed to be prepared, copied, and parts learned by the participants.

Props would be kept to the minimum, basically nothing more than making use of what was available in the lecture hall: a bench, a lecture table, and some chairs. But it was the tenth anniversary of the founding of the institute as well as the hundredth anniversary of Goethe's death. The skit should be worthy of the occasion.

Three astrophysicists as the Archangels.

With the meeting drawing to a close, the participants gathered in the institute's first-floor lecture hall for the performance. Goethe's *Faust* has a Prologue, in which the three Archangels, Raphael, Michael, and Gabriel, laud the Lord for the creation of the heavens. Mephistopheles then appears, sarcastically jibing the Lord for his seriousness:

> My pathos soon thy laughter would awake,
> Hadst thou the laughing mood not long forsworn.
>
> *Faust,* Prologue, 35–36

In the skit, the Archangels have become three astrophysicists, sitting together behind a lecture table at the front of the hall. They explain how physics has led them to understanding the stars and the cosmos, concluding in unison with a paean to Bohr/the Lord that includes a sly reference to the well-known difficulty in deciphering Bohr's writings:

> *This vision fills us with elation*
> *(Though none of us can understand).*
> *As on the Day of Publication*
> *The brilliant Works are strange and grand.*

The audience noticed during the speech a cotton sheet covering a large draped figure sitting on a stool next to the table. It was now lifted, revealing Felix Bloch made up to uncannily resemble Bohr.

Perhaps the honor of playing Bohr should have gone to Leon Rosenfeld, a Belgian who arrived in Copenhagen in 1931 to replace the departing Oscar Klein as Bohr's assistant. Rosenfeld remained Bohr's closest collaborator through the 1930s, but he was short, balding, and a little pudgy whereas Bloch, like Bohr, was powerfully built and athletic. Rosenfeld looked much more like Pauli, and playing Pauli was how the audience saw him. Disguised with horns and tail, he came running into the hall and jumped up on the table. Crouching, he addressed the Lord and then the audience:

> *Since you,* O Lord, *yourself have now seen fit*
> *To visit us and see how each behaves,*
> *And since it seems you favor me a bit,*
> *Well—now you see me here among the slaves.*

The audience of *slaves* roared. So did Bohr. Unlike the Lord who suffers Mephistopheles' reprimand in Goethe's work, laughing was definitely not a habit Bohr had forsworn.

Bloch and Rosenfeld, the Lord and Mephisto, then began to formulate their wager about the possession of Faust's soul:

> THE LORD
> *You know this* Ehrenfest? . . .
>
> MEPHISTO
> *The Critic?*

Having been summoned, Ehrenfest/Faust appears and begins to lament how little physics he has understood despite his prolonged efforts. In addition he suffered as

> All *doubts assail me; so does* every *scruple;*
> *And Pauli as the Devil himself I fear.*

In Goethe's masterpiece, Faust, transformed by a potion, soon meets young Gretchen, falls in love with her, and seduces her. This liaison leads to a scene where Gretchen, alone in her room and sitting at a spinning wheel, tells of her love for Faust and her fear that he has abandoned her; the grief she feels is driving her mad.

The neutrino/Gretchen singing her song.

Gretchen has become the neutrino in the parody's centerpiece. To the tune of Franz Schubert's "Gretchen at the Spinning Wheel," a young woman sang, now paraphrasing Goethe's lines, that

> *My Mass is zero,*
> *My Charge is the same.*
> *You are my hero,*
> Neutrino*'s my name.*
>
> *I am your fate*
> *And I'm your key.*
> *Closed is the gate*
> *For lack of me.*
>
> *Beta-rays throng*
> *With me to pair.*
> *The N-spin's wrong*
> *If* I'm *not there.*

There is a bit of a swindle in casting Gretchen as the neutrino, which was Pauli's love and creation: Gretchen is Faust's love, not Mephistopheles'. On the other hand, it was dramatically correct as well as irresistible to identify Pauli—the well-known *Scourge of God*—with Mephistopheles. Incidentally, Gretchen's part was played by a friend of Delbrück's, a young Danish woman named Ellen Tvede. A few months later he introduced her to his old Göttingen comrade Viki Weisskopf, newly arrived at the Bohr institute as a Rockefeller Fellow. Ellen and Viki were married in 1934.

Delbrück's Dilemma

THE FIRST PART of the skit had been easy to write once Delbrück thought of centering it on the neutrino, but he still faced two problems. The first was how to deal with Einstein. He never came to Copenhagen and, despite his great friendship with Bohr, did not take much of an interest in the subjects the young scientists there were working on, such as nuclear physics. That didn't mean Einstein could be dismissed or ignored. He certainly was the twentieth century's greatest physicist, and with the possible exception of Newton, the greatest of all time. Despite the misgivings Delbrück and his cohorts had about the direction of Einstein's research, the "Copenhagen Faust" had to include him. But how could Delbrück do it in a way that reflected his own generation's disinterested view of Einstein's attempts to derive a unified field theory of gravity and electromagnetism? He thought hard and found a way.

In *Part I* of Goethe's drama, Faust and Mephistopheles go into Leipzig's Auerbachs Keller, an ancient beer and wine cellar still in existence; they meet there a group of drinkers. To amuse the revelers, Mephistopheles sings a song about a king who has a giant flea (four lines of it are the epigraph to chapter 9). The flea has more fleas and, by order of the king, nobody is allowed to touch them. The suffering court's plight is described:

> The gentlemen and ladies
> At court were sore distressed;
> The queen and all her maidens
> Were bitten by the pest,
> And yet they dared not scratch them

Or chase the fleas away.
If we are bit, we catch them,
And crack without delay.

(*Faust, Part I,* 1880–87)

Einstein/the King leading his pet flea.

The assembled cheer, drink, and toast Mephistopheles.

In Delbrück's version, Einstein is the king, an ironic nod of respect since his new theories are compared to the fleas that torture the king's court. In view of the fact that Einstein was in the United States in the spring of 1932, Max changed the Auerbachs Keller into an American speakeasy. Entering it, Mephisto introduces himself, saying,

Can no one laugh? Will no one drink?
I'll *teach you physics in a wink. . . .*
Shame on you, sitting in a daze
When as a rule, you are all ablaze!

They all then sit back and listen, as Mephisto sings,

Half-naked, fleas came pouring
From Berlin's joy and pride,
Named by the unadoring:
"Field Theories—Unified."

So the fleas had become Einstein's unified field theories of electromagnetism and gravity, no more than a nuisance to the *unadoring* King's subjects. It is also tragically true that by this point Einstein was no longer Berlin's pride and joy, if he had ever been. By the following year he wouldn't even be welcome in Germany.

The second problem Delbrück had to deal with was incorporating into the parody something extraordinarily important that had happened two months earlier, an experimental discovery that was bound to change everything they thought about the atomic nucleus. Working at the Cavendish Laboratory, James Chadwick had discovered the neutron, the proton's neutral counterpart within the nucleus. Delbrück didn't want the neutron to be the drama's central theme, but leaving it out was not an option. It was, after all, what they had been talking about all week, and the skit had to include the novelty that was shaking their world. But how could the neutron and its discoverer fit into the Faust story?

Chadwick/Wagner—The Ideal Experimentalist is balancing a black ball on his finger.

Delbrück again found a way. In Goethe's drama, Faust has an assistant named Wagner. We first meet him as he interrupts Faust's initial reverie. Lamenting his own lack of accomplishment, Wagner seems no happier than Faust. But in *Part II* he returns to the stage, rejoicing about having performed a miraculous experiment, the production in a test tube of Homunculus, a small artificial man.

In the Faust parody Wagner became Chadwick/Wagner. Labeled as the Ideal Experimentalist, he enters, quickly announcing his discovery:

> *The* Neutron *has come to be.*
> *Loaded with mass is he.*
> *Of Charge, forever free.*
> *Pauli, do you agree?*

To which Pauli/Mephisto can only reply,

> *That which experiment has found—*
> *Though theory had no part in—*
> *Is always reckoned more than sound*
> *To put your mind and heart in.*
> *Good luck you heavyweight Ersatz—*
> *We welcome you with pleasure!*
> *But passion ever spins our plots,*
> *And Gretchen is my treasure!*

The lines summarize nicely Pauli's attitudes—his belief that theorists should be guided by experimental results, his acceptance of the neutron, and his lasting affection for the neutrino/Gretchen, his own creation.

The neutrino did remain Pauli's treasure. When in 1956 he received notice that it had definitively been observed, he sent the discoverers a telegram: "Thanks for the message. Everything comes to him who knows how to wait."

Since 1956 three Nobel Prizes in Physics have been awarded for experiments centered on neutrinos: one for the first experimental observation, a second for the determination that another type of neutrino exists, and a third for the observation of neutrinos emitted by the sun's core. However, to my knowledge, there has never been the slightest practical application of neutrinos, either for good or for evil. They interact so weakly with matter and are so damnably hard to detect that nobody has found any use for them other than to advance our understanding of the basic laws of nature.

The neutron, on the other hand, "loaded with mass" as it is, is a different story. Its detection reasserted the importance of experiments and marked an important transition in physics, from concentrating on the atom's electrons to studying its nucleus. And the applications of that research have certainly changed the world in ways the scientists could not imagine in 1932. In their pursuit of knowledge they had uncovered a truth with implicit powers for both good and evil.

In the spirit of scientific discovery, the physicists had found something that was new and exciting, one that lifted the veil off mysteries they had long sought to uncover. Their quest was to be applauded. The devil, however, was in the details. While their findings would eventually yield sources of energy that could supplant coal and oil, those same findings would soon give birth to apocalyptic weapons. Unbeknownst to the scientists, they were setting out on a path that would lead to the striking of Faustian bargains. It was not long before they discovered their magnificent discovery's heavy price.

Chapter 13

The Miracle Year

WAGNER *(AT THE FURNACE)*
Soundeth the bell, the fearful clang
Thrills through these sooty walls; no more
Upon fulfilment waits the pang
Of hope or fear;—suspense is o'er;
The darkness begins to clear,
Within the inmost phial glows
Radiance, like living coal, that throws,
As from a splendid carbuncle, its rays.

Goethe, *Faust, Part II,* act 2, 254–61

The Discovery of the Neutron

JAMES CHADWICK'S daily custom, as second-in-command at the Cavendish, was to report in the morning to Rutherford on interesting developments of the previous twenty-four hours, whether in the Cavendish or elsewhere. On a cold Cambridge morning in February 1932, there was something he very much wanted to discuss with Rutherford.

He had read that morning a paper by Irène Curie and her husband, Frédéric Joliot, both accomplished experimentalists working in the Paris laboratory that was still headed by Marie Curie, Irène's mother. Entitled "The Emission of Protons of High Velocity from Hydrogenous Materials by Very Penetrating Gamma Rays," it had been submitted to the

Comptes Rendus de l'Académie des Sciences (Proceedings of the French Academy of Sciences) two weeks earlier and published immediately. Chadwick suspected the experiment was right, but was skeptical of its conclusions.

Walter Bothe, a distinguished Berlin experimentalist, had found in the late 1920s that beryllium nuclei, after alpha particle bombardment, emit a very penetrating radiation he presumed were gamma rays, beams of photons. Following up on Bothe's result and with access to the Curie laboratory's strong source of alpha particles, Curie and Joliot began bombarding beryllium nuclei as Bothe had done and then exposing paraffin wax, a material rich in hydrogen, to the radiation emitted by the beryllium nuclei. They discovered that the supposed photons knocked hydrogen nuclei (protons) out of the paraffin at a truly stupendous rate, three million times more abundantly than they expected. This is what they were reporting.

Chadwick thought it highly unlikely that gamma rays (photons) could achieve the effect Curie and Joliot measured. Many experiments had been done in which gamma rays set electrons in motion. However, inducing significant movement in a proton, almost two thousand times as massive as an electron, seemed scarcely credible. Metaphorically, it is one thing to set a tricycle rolling with a firm push and quite another to similarly move a truck.

On the other hand, if the original alpha rays knocked something substantial out of the beryllium nucleus, it would be rather easy for that massive object to in turn propel a proton from the target. It just so happened that *something substantial* had been on Rutherford and Chadwick's mind for a long time.

Rutherford had been emphasizing since 1920 the necessity for particles other than electrons and protons to be inside the nucleus. How else could one imagine large nuclei being assembled? He thought this something else—he called it a neutron—was an electrically neutral particle with roughly the same mass as a proton, perhaps simply a proton bound tightly to the much lighter electron. His longtime collaborator Chadwick remembered their frequent discussions on the subject and their frustration at not finding any evidence for this massive particle, presumably so hard to discover because it was electrically neutral. As years passed without finding it, the neutron's elusiveness became a source of constant frustration for the two scientists.

Chadwick now asked himself what if Bothe, Curie, and Joliot had been wrong? What if those rays emitted by the beryllium nuclei were really neutrons? It would be an experiment that would change everybody's view of the nucleus.

At 11 a.m. Chadwick went in to see Rutherford, as usual. He remembered clearly Rutherford's reaction to the news of the Paris findings: "As I told him about the Curie-Joliot observation and their views on it, I saw his growing amazement; and finally he burst out, 'I don't believe it.' Such an impatient remark was entirely out of character, and in all my long association with him, I recall no similar occasion." Galvanized by Rutherford, Chadwick set to work.

Chadwick is often described as birdlike in appearance, with round tortoiseshell eyeglasses on a thin nose and a long, rather severe face. Rutherford was still the driving force at the Cavendish, but the forty-one-year-old Chadwick was the man who presided over the day-to-day workings of the laboratory. At heart he was still a passionate experimentalist. The spirit that had driven him to do physics experiments even while half-starved in a World War I German internment camp was very much alive.

Chadwick's reaction to the news was to begin trying to prove or disprove the Paris conclusion. He quickly verified Curie-Joliot's results and then set about systematically bombarding a series of other materials with the beryllium emitted radiation. His and Rutherford's intuition had been correct. The conclusion was inescapable. Chadwick had discovered the neutron.

On February 17 he sat down at his desk to write up his findings. He had been working almost continuously for ten days, averaging at most three hours of sleep a night. But he had now finished. His report to *Nature*, entitled "Possible Existence of a Neutron," concludes this way:

> If we suppose that the radiation is not a gamma radiation, but consists of particles of mass very nearly equal to that of the proton, all the difficulties associated with the collisions disappear, both with regard to their frequency and to the energy transfer to the different masses. In order to explain the great penetrating power of the radiation we must further assume that the particle has no net charge . . . We may suppose it to be the "neutron" discussed by Rutherford in his Bakerian lecture of 1920.

The hunt was finally over. After more than ten years, Chadwick had captured his prey.

His good friend Pyotr Kapitza took him to dinner that night. Very social and a brilliant physicist, Kapitza organized impromptu, fun-filled Cambridge after-dinner physics discussions that came to be known as meetings of the Kapitza Club. Chadwick happened to be the speaker on the evening his experiment concluded. The novelist-scientist C. P. Snow, present at the occasion, remembered Chadwick giving a brief presentation about his great discovery and then, with weariness showing in his face, saying, "Now I want to be chloroformed and put to bed for a fortnight."

Chadwick had worked around the clock, knowing the Paris group might realize their error at any time and beat him to the result. It was a sweet victory for the Cavendish. Chadwick quickly received the Nobel Prize in Physics. It seems that Rutherford insisted that it not be shared with Curie and Joliot, declaring, "For the neutron, to Chadwick alone, the Joliots are so clever that they soon will deserve it for something else." It turned out that this prediction was right as well. Chadwick alone received the 1935 physics prize for his discovery of the neutron. The Joliots went to Stockholm the same year to accept the Nobel Prize in Chemistry for their 1934 detection of induced radioactivity.

Ushering in the era of intense research in nuclear physics, Chadwick's achievement was almost as important as Rutherford's original 1913 realization that the atom had a nucleus, but the effect on physicists was very different this time. The 1913 result had been a total surprise, a bolt out of the blue, leaving the community and even Rutherford undecided on how to proceed. By contrast, the news of the neutron's discovery, although not generally anticipated, fell on fertile ground, solving almost immediately a number of puzzles and providing fodder for new ones.

Copenhagen and the Neutron

THE EXISTENCE OF this new particle was the most exciting topic at the Bohr institute's spring meeting, held two months later in 1932. The Copenhagen gathering was the ideal place to consider all the facets of a new experimental result with unclear implications. With no set a priori agenda for the week's discussions, the group could devote as much time as they wanted to what might be the experiment's consequences, trading ideas with one another on how to proceed. Some of the issues raised that week were new, but others, contemplated for years, needed to be reexamined under this new light.

The outstanding old problems were the unexplained mass of nuclei and the source of the mysterious force that holds nuclei together. Could seven neutrons in the nitrogen nucleus be the source of its missing mass? Could this be confirmed? If it was the case, the so-called wrong-statistics problem, the question of why a nitrogen nucleus acts as if it has an even number of particles, not an odd number, might also be settled, because 7 + 7 = 14. Was this true? Were neutrons responsible for holding large nuclei together? If that was the case, how did they do it?

At the skit's end, Pauli/Mephisto reaffirmed the theorists' faith in experiment and wished the neutron well:

That which experiment has found—
Though theory had no part in—
Is always reckoned more than sound
To put your mind and heart in.
Good luck you heavyweight Ersatz—

But how could the "heavyweight Ersatz" be compatible with Gretchen? How did the neutron and the neutrino fit together in the nucleus? How did beta rays (electrons) come out of the nucleus? When discussions during the meeting reached a point of such complexity, Bohr might lighten the mood by repeating one of his favorite sayings, "A great truth is a truth whose opposite is also a great truth," egging his young followers to think boldly.

The physicists congregating in Copenhagen that week decided that understanding the neutron's behavior required determining if it was on the same footing as the electron and proton or if it was simply an electron and a proton bound together. Reluctant to introduce more new particles, the participants seemed to favor the latter alternative. As Chadwick said, "The possibility that the neutron might be elementary has little to recommend it at present, except the possibility of explaining the statistics of such nuclei as nitrogen-14."

Pragmatic and daring as usual, Heisenberg acted while others pondered whether the neutron was *elementary* or not. Within months of the meeting's end, he formulated an elegant quantum mechanical theory of the forces between neutrons and protons that did not depend on knowing the answer to that question. This quickly became the basis for all future work along these lines, proving once again the wisdom of Peierls's assessment of Heisenberg: "When he was faced with a problem, he would almost always intuitively know what the answer would be and then look for a mathematical method likely to give him that answer."

The burgeoning field of nuclear physics continued to gain speed. By 1934 the evidence clearly indicated that neutrons and protons were the nucleus's building blocks, with the neutron every bit as fundamental as the electron and the proton. But this still did not explain beta decay. How could electrons come out of the nucleus if they weren't inside to begin with?

Fermi offered a daring and elegant solution in late 1933. A new force of nature was required, conceptually on the same level as the two known forces, gravitational and electric. Termed the *weak force*, it differed from the other two by changing particles' identities; as envisioned by Fermi, its action would turn an electron into a neutrino and a proton into a neutron. By a slight extension, it could even lead to a neutron disappearing altogether, replaced by a proton, an electron, and a neutrino. This new force's manifestations were almost always hidden, rising to clear visibility in only a single instance, nuclear decay by the emission of beta rays.

In one dramatic stroke, Fermi had apparently solved the major problem of beta decay, namely, How do electrons come out of the nucleus if they aren't there to begin with? We must think of a nucleus as comprising neutrons and protons, but when the weak force acts, a neutron inside the nucleus is replaced by a proton while the accompanying energetic electron and neutrino, created at that instant, immediately escape the nucleus they were created in.

The notion that elementary particles could change their identity was even more revolutionary than the existence of a new force. Yet, once proposed, it seemed almost inevitable to Fermi and others. My uncle Emilio remembered sitting with a few friends in a hotel room in the Dolomites after a day of end-of-the-year skiing in 1933 as Fermi explained his revelation. Fully aware of its importance, he told his dumbstruck friends in an unemotional way that this was likely to be the most important piece of work he had done or would ever do in his life.

Ever cognizant of the details of experimental work, Fermi also studied the shape of the curve describing the measured electron energies and concluded, in his own words, that the neutrino mass was "equal to zero, or in any case, small compared to the electron mass." How to measure that small neutrino mass was the central issue of the neutrino conference I was attending in Munich seventy years later. One always hears stories about the rapid pace of science, but some things take a little longer.

Fermi quickly wrote a short paper on the subject of his proposed solution for understanding beta decay and sent it off for rapid publication to *Nature*. The journal, suspicious of new forces and undetected particles, rejected the paper with an accompanying editorial comment to its author: "The conjectures are too removed from physical reality." Theoretical physicists, quickly accepting the new proposal, did not share this opinion, and in the ensuing years Fermi's theory was shown to be almost entirely right. It is now one of the cornerstones of elementary particle physics, and *Nature*'s refusal is cited as a classic case of a journal being overly timid in its selections.

I have some sympathy, however, for *Nature*'s editors. The speed and complexity of the new physics was quite overwhelming for all but the experts. Two years earlier the world had consisted of electrons and protons held together by electric forces. Now there were supposedly two new particles, neutrons and neutrinos, an unknown mechanism holding neutrons and protons together and yet another unknown mechanism allowing neutrons to decay. To make matters even more complicated, one of the new particles had originally also been called neutron and was just now getting a name change to neutrino.

Yet all those ideas were correct. More than a dozen physicists have received Nobel Prizes in the past half century for proving that the weak force violates parity, for showing the details of how the force is mediated, for proving the existence of neutrinos, for showing how they are produced in the sun's nuclear reactions, and more. But all that work had its origins in the early 1930s ideas of Pauli and Fermi and further back in the marvelous experiments of Rutherford, Chadwick, Ellis, Meitner, and their co-workers. Such is the weave of the scientific fabric.

The Miracle Year

THE NEUTRON'S DISCOVERY came as a surprise, but the year's most sensational result, and conceptually the most radical, was not announced until after the Copenhagen meeting. In the late summer of 1932, Carl Anderson, a young American working at the California Institute of Technology, presented incontrovertible evidence for the existence of a particle with the same mass as an electron but opposite electric charge. There was no possibility of its being a proton, almost two thousand times as massive as an electron. Without knowing about Dirac's theory, still regarded as an abstruse mathematical concept, Anderson had discovered the antielectron, soon given the name of *positron*.

Once the result was known, many other experimentalists realized that they had seen positrons in their own apparatus but had dismissed their findings as erroneous or had misinterpreted them—the track of a positron moving in one direction could be mistaken for that of an electron moving in the opposite direction. Two prominent Cambridge experimentalists had gone so far as to discuss positrons' possible existence with Dirac after seeing some evidence for them. Yet they hesitated to publish, reluctant to accept the seemingly bizarre notion that antimatter existed. Even after performing further experiments that supported Dirac's hypothesis, they concluded their published paper with only a timid acknowledgment of the particles' existence: "When the behavior of the positive electrons have been investigated in more detail, it will be possible to test these predictions of Dirac's theory. There appears to be no evidence against its validity."

The authors of those words were Patrick Blackett and Giuseppe Occhialini. Occhialini, spending two years in Cambridge on leave from his post in Florence, was a good friend of my parents, a witness at their wedding. Exuberant and principled, an adventurous soul, Occhialini left Italy before World War II, repelled by the Fascist government in power. Seeking refuge in Brazil and forced for a while to abandon physics, he earned a living as a mountain guide in the Brazilian highlands. Returning to Europe before the war's end to fight with the British, he finally made his way back to a liberated Italy. This is all by way of saying Occhialini was not a timid man. Yet even he hesitated to assert more than "There appears to be no evidence . . ."

Others were skeptical as well because Dirac's theory was such a break with the past. As Heisenberg said, "The discovery of antimatter was perhaps the biggest jump of all the big jumps in our century. It was a discovery of utmost importance because it changed our whole picture of matter." At first Bohr did not believe the results of Anderson's experiment, his hesitation provoked by the seeming unlikelihood of Dirac's infinite sea of negative energy states with holes in it. This construct was so unpalatable to Bohr and to Pauli that they separately wrote Dirac stating they could not accept it even if antielectrons were shown to exist.

Lise Meitner decided to test the conclusions of Anderson and the Cambridge duo. Their experiments had been based on analyzing cosmic-ray-initiated reactions, *cosmic ray* being a generic term for radiation coming from outer space. There was no a priori reason for assuming that photons coming from the cosmos should be any different from photons produced in a laboratory, but she knew there is no real substitute for the kind of control and reproducibility one has in a laboratory experiment. Setting to work, she quickly confirmed and then extended the others' results.

The first verification of a threshold for photons' ability to produce antimatter was an important aspect of Meitner's findings. Einstein's famous $E = mc^2$ reflects the energy content associated with mass, but physicists had not contemplated the notion that a particle's mass might disappear altogether, reappearing as energy. Yet that is exactly what happens when a particle meets its antiparticle. The annihilation of an electron and a positron can lead to only photons appearing; energy has been conserved, but the mass of the partners has completely vanished in the encounter. Conversely, if photons are energetic enough, two of them can transform themselves into an electron-positron pair, effectively creating mass out of pure energy. In order to do so the photons' energy must at least equal the mc^2 of the electron and of the positron. Meitner's experiment confirmed this prediction. *Pair production* now became a new term in physicists' lexicon.

As always, theory stimulates experiment and experiment in turn stimulates theory. Meitner's assistant, or as he had called himself in his 1932 letter to Bohr, her "family-theorist," was also at work. Delbrück began to think of how antimatter might affect photons. It was known that the presence of other photons would not change the path of a photon, but if those other photons could convert into electron-positron pairs, even for an instant, the picture changed. Meitner's paper has a very interesting appendix on the subject. Explaining what we now call Delbrück scattering, this turned out to be his best-known contribution to theoretical physics, one still studied today.

Electron-positron creation solved the Klein paradox of apparent number nonconservation, the problem that had concerned Bohr so deeply. He had made his proposal of energy nonconservation on the scale of the nucleus in 1929, thinking it might provide a common solution to four separate problems troubling the physics community. By the end of 1933 it was clear that the answer to the other three riddles also lay elsewhere. The so-called wrong statistics of nitrogen was corrected by neutrons, missing energy in nuclear beta decay was explained by neutrinos, and finally, anomalous energy production in stars was no longer anomalous once physicists understood the details of nuclear reactions.

In the end, energy retained its significance as the fundamental parameter of nature, its conservation remaining the bedrock of all interactions. Bohr's official concession came in a short letter to *Nature* published in June 1936. Entitled "Conservation Laws in Quantum Theory," it concludes with these words:

> Finally, it may be remarked that the grounds for serious doubt as regards the strict validity of conservation laws in the problem of the emission of beta rays from atomic nuclei are now largely removed by the suggestive agreement between the rapidly increasing experimental evidence regarding beta ray phenomena and the consequences of the neutrino hypothesis by Pauli, so remarkably developed in Fermi's theory.

Discovering the positron and the neutron were the two most surprising and acclaimed experimental achievements of a momentous year. The first confirmed the essential correctness of Dirac's theory while the second launched the field of nuclear physics. Hans Bethe, a man who won a Nobel Prize for elucidating the details of the nuclear physics reactions that power the sun's core, called the period before 1932 "the prehistory of nuclear physics, and from 1932 on, the history of nuclear physics." The neutron's discovery was the dividing line, and 1932 was known as the Miracle Year.

The preceding years' greatest advance had been the development of quantum mechanics. But the pendulum swung away from theory in 1932, spurred by groundbreaking experiments, numerous and profound. That's not to say that there weren't also extraordinary steps forward in theory during the 1932–34 period, but they were usually driven by the discovery of new and unexpected data. Experiments were now guiding progress. One of the most important ones was brought to its conclusion in April 1932, just as the meeting was taking place in Copenhagen.

John Cockcroft and Ernest Walton at the Cavendish had finally achieved, after more than three years of building and planning, what Rutherford had told them to do after his conversation with Gamow: "Build me a one-million electron-volt accelerator; we will crack the lithium nucleus without any trouble." Using protons as projectiles, they had done exactly that.

Bohr wrote Rutherford when informed of Cockcroft and Walton's success: "Progress in the field of nuclear constitution is at the moment really so rapid, that one wonders what the next post will bring." He didn't have to wait long. Shortly after Rutherford's news, Otto Stern, the Hamburg experimentalist and master of the ultradifficult experiment, told Bohr that he had managed to measure the proton's so-called magnetic moment, the indicator of how a proton would move when placed in a magnetic field. If the proton, as expected, was the same sort of particle as the electron, only with more mass and opposite charge, Dirac's equation predicted the answer.

Pauli, Stern's old friend from Hamburg days in the 1920s, knowing how much Stern enjoyed doing hard experiments, told him to go ahead and try to make the measurement, although Pauli claimed everybody already knew the answer. Confounding expectations, another bombshell exploded. Stern's result was three times larger than the prediction; there was more to the proton than Pauli or anybody else had suspected. It was the first hint that the proton and the neutron had substructures of their own.

The challenge at the beginning of the century had been to prove that matter was made of atoms. Then Rutherford had shown that atoms were nuclei surrounded by electrons. Nuclei were then seen to be protons and neutrons, and these had now been shown to have a structure of their own. It seemed to be a marathon of discoveries with no foreseeable end.

Big Science Is Born

ONE THING WAS CLEAR: bigger would be better. Paradoxically, probing the nucleus at ever-smaller lengths requires projectiles of ever-greater energy. Taking this into account, a new experimental program for the future was now envisioned. Paraphrasing Rutherford, its message would be, "Build me a one-*billion* electron-volt accelerator; we will crack the proton without any trouble." But building such a machine would require a different approach. Rutherford's way, as head of the Cavendish, had always emphasized self-reliance and frugality. You built, assembled, and conducted your own experiment, perhaps in partnership with somebody else; occasionally three people worked together, but no more. No assignment of subtasks, no special technicians, and no team collaborations were needed. By and large this was also how other experimentalists, such as Stern and Meitner, did their research. But the old way was no longer possible if physicists were going to build and operate a one-billion-volt accelerator.

This new style of physics experimentation also made its debut in 1932. California's Ernest Lawrence, using the first prototype of his new machine, the cyclotron, replicated Cockcroft and Walton's results in September of that year. Encouraged, Lawrence was soon talking of building larger and improved cyclotrons. What had started as a tabletop experiment quickly became a large enterprise, involving specialized technicians and a permanent staff, all at considerable expense for construction and maintenance.

Driven by advances in technology, the lead in this new venture passed to the United States and would remain there for the next fifty years; atom smashing's future belonged to Berkeley, not to the Cavendish, and with that an era ended. By the close of the decade, the United States had a dozen cyclotrons and Europe had five. Machines became bigger and bigger, in a process that has now marched forward for almost seventy-five years. It is now a world effort.

Having spent three years at CERN, the organization that hosts the largest heir to those early machines, I have come to appreciate what an extraordinary place it is. Started in 1954 by a consortium of European nations with the intent of bringing leading subatomic research back to Europe, it is located at the border between France and Switzerland, on the outskirts of Geneva. Presently housing a staff of some three thousand and operating at an annual cost of close to a billion dollars, CERN's core is still European, but visitors from around the world participate in its research. It has become a mini-U.N. of science, a remarkable example of international collaboration cutting across national boundaries.

The facility that dominated CERN's activities in the 1990s was known as LEP, the Large Electron-Positron collider. In it, beams of electrons and positrons were accelerated to velocities close to that of light by powerful magnetic fields. The beams circled in opposite directions more than ten thousand times a second in a twenty-seven-kilometer underground tunnel, crossing from Switzerland to France to Switzerland, back to France and then back again to Switzerland. To reduce beam collisions with air molecules, the tunnel was kept at a high vacuum, less than a billionth of atmospheric pressure. The beams were directed to partly collide in four giant collision detectors, Aleph, Delphi, L3, and Opal, each run by an international collaboration of more than three hundred physicists.

Sixty years earlier, physicists had been amazed to discover the positron. Now beams of them were being used as a research tool. The experiments were so sensitive that even the water level of Lake Geneva had to be taken into account. Though ten kilometers away, the lake's seasonal variations shifted the pressure on its shore enough to move the beam's location by a fraction of a millimeter. Such a disturbance, as well as the passing of trains and the changing orbit of the moon, had to be taken into account to gain the desired maximum accuracy. Meanwhile the sheer volume of data obtained at Aleph, Delphi, L3, and Opal was so large that examining experimental results stretched CERN's giant computational facilities to their limits. To coordinate the efforts of physicists around the world in analyzing that data, CERN developed a tool that has come to be known by its WWW acronym, the World Wide Web.

It had been twenty years from Rutherford's discovery of the nucleus to Chadwick's of the neutron, and it would be another twenty until physicists began to understand what the proton and neutron were made of, and yet another twenty for them to realize that the precise answer is three quarks. It would require twenty more before many within the scientific community began to think of those quarks as superstring vibrations. Right now we are waiting for the thousand billion electron-volt accelerator. The LHC (Large Hadron Collider), LEP's CERN successor, expected to be operational within the next few years, may provide answers to a number of remaining questions.

The machines have come to cost billions, whether dollars, Swiss francs, or euros, and experimental groups have grown from a single individual or two to many hundreds. National boundaries have been breached to finance the machines and run the experiments. Big science was born in 1932.

The Hammer and the Needle

NUCLEAR PHYSICS had advanced by studying the bombardment of atoms with beams of hydrogen ions (protons) or of helium ions (alpha particles). After twenty years of research, the technique pioneered by Rutherford had culminated in Cockcroft and Walton's experiment, but there was now a new type of projectile, the neutron. Having to overcome the electric repulsion between either protons or alpha particles and similarly charged nuclei had always been a limitation on nuclear physics experiments. Neutrons, electrically neutral, suffered no such problem.

On the other hand, seen from another point of view, this neutrality was a disadvantage. Because neutrons are impervious to electrical forces, those forces could not be used to accelerate them. Neutrons are perfect nuclear probes if you are satisfied with relatively low-energy projectiles, but should you have wanted in 1932 to break nuclei into pieces, you needed protons or alpha particles, accelerated until they became ultra-energetic. In other words neutrons are ideal for tickling nuclei whereas accelerated protons or alpha particles are needed to smash them. One is a needle and the other a hammer.

While the hammer was leading the way to big science in 1932, the needle was having its own impact. Although 1932 had ushered in the era of nuclear physics, few scientists thought at the time of using the knowledge for practical applications other than perhaps in the treatment of certain types of cancer. Indeed Lord Rutherford, blunt as usual, warned the British Association for the Advancement of Science in late 1933, "To those who look for sources of power in atomic transmutations—such expectations are the merest moonshine."

Those expectations changed with the 1934 discovery by Curie and Joliot that some substances become radioactive when bombarded with alpha particles. Radioactivity was a known source of energy; if it could be induced at will, there was hope of finding new "sources of power." Very soon thereafter, using slow neutrons as bombarding particles, Fermi's Rome group observed the same induced radioactivity that Curie and Joliot had seen, but at a far greater rate. This was a major step toward the building of nuclear reactors. As the needle began to be employed, energy generation by nuclear physics techniques, contrary to Rutherford's statement, was looking increasingly possible.

The Rome group's research had started with a remarkable transition Fermi made in 1934, from theoretical to experimental physicist. Close inspection shows the change was not abrupt, or unplanned, and not permanent. For the rest of his life Fermi went back and forth, unique as both a front-rank theorist and experimentalist. Most theoretical physicists follow closely the results of experiments, but Fermi was different in actually wanting to perform them.

His early successes had all been as a theoretical physicist, but once established in Rome, Fermi had begun to plan a modern experimental program in the antiquated physics department of its university. The first step was to obtain fellowships for his junior collaborators to go abroad and learn up-to-date experimental techniques. Off they went, to Pasadena, Amsterdam, Leipzig, Hamburg, Berlin; then they all came back to Rome. Franco Rasetti's two stays, the second one after the neutron's discovery, in Meitner's Berlin laboratory were particularly important. He was the senior member of the young group around Fermi, the one closest to him, having been Fermi's friend and companion since university days. While Fermi was given the nickname of the Pope, Rasetti was known as the Vicar Cardinal.

Using neutrons as projectiles was what Fermi had been waiting for: a brand-new field of nuclear physics. By early 1934 the two Italians were building measuring devices for use in their anticipated nuclear research. Success with their new equipment was almost immediate. Its significance was recognized by Rutherford. He sent Fermi a letter in April 1934 expressing his interest in the work, applauding the initial results, and then facetiously adding, "I congratulate you on your successful escape from the sphere of theoretical physics." Fermi now sent two letters describing the new measurements to *Nature* for publication. They were accepted immediately, and in July Rutherford forwarded a detailed paper written by the Rome group to the *Proceedings of the Royal Society*. Earlier this same year *Nature* had rejected for publication Fermi's work on the weak interaction as being "too speculative," but the contents of the new publications were deemed to be facts, not speculation.

To appreciate how fast the physics world was changing, remember that Chadwick discovered the neutron in February 1932, and two years later Fermi was using neutrons as projectiles to study nuclear reactions. As often happens in science, yesterday's discovery is today's tool.

Research with the needle had another far-reaching consequence. If a neutron entering a nucleus were to alter that nucleus in such a way that it released two neutrons, these two could in turn perforate two other nuclei and free four neutrons. Two to four, four to eight, eight to sixteen, . . . If energy was released every time the needle was inserted and if the cascade were to happen rapidly enough so that the released energy did not stop the sequence by blowing the target apart, one could have a chain reaction that quickly involved almost all the nuclei in the target. The effect of tapping into those enormous reservoirs of energy would be unimaginable; the explosive power of thousands of tons of dynamite could be obtained by using just a few pounds of fissioning material. The bomb would be small enough to be carried in a truck or an airplane or perhaps even some day in a suitcase.

It seemed like a very far-fetched notion, but when it occurred to a cherubic-faced Hungarian physicist named Leo Szilard, he was alarmed. It was September 1933. Szilard recognized the potential of a chain reaction for building weapons. Of course their construction would require an immense technological effort. The outcome would be uncertain, the expenditures of money enormous. However, having fled from Germany to England only a few months earlier, Szilard was keenly attuned to even a small prospect of advances in weaponry.

He was well aware of the highly developed state of German science and of the increasingly worrisome political climate taking hold there. Szilard had spent most of the 1920s in Berlin, first as a student and then as a lecturer at the university, but after Adolf Hitler had been appointed

German chancellor on January 30, 1933, he began to plan his departure. Hitler's control over the German government increased after the burning of the Reichstag, the parliament building, on February 27. In late March 1933, the new chancellor was essentially granted dictatorial control of the country by the passage of the so-called Enabling Act. Massive persecution of Jews began almost immediately, and Szilard, himself a Jew, left Germany at once.

Szilard's Berlin friend and collaborator Albert Einstein also did so. More than a decade of anti-Semitic attacks directed against Einstein left him with a dread of what might be Germany's future. In early March 1933, on a prolonged visit to the United States, he gave an interview to a New York newspaper making public his decision not to go back to Germany, a country without "civil liberty, tolerance and equality before the law." Sailing to Europe, he disembarked at Antwerp and drove from there straight to the German embassy in Brussels, where he formally surrendered his passport and renounced German citizenship. Soon thereafter he returned to the United States, never setting foot in Europe again.

Chapter 14

Ehrenfest's End

WAGNER
Oh God! How long is art,
Our life how short! With earnest zeal
Still as I ply the critic's task, I feel
A strange oppression both of head and heart.

Goethe, *Faust, Part I,* 210–13

THE DEPARTURE of his dear friend Einstein was a bitter blow for Ehrenfest. Anxious, saddened, and depressed by Hitler's ascent to power and the decline of all the German institutions he valued, Ehrenfest watched, in a state of semiparalysis, his heroes disappear. Lorentz had died, Planck was increasingly being placed in the ambiguous situation of defending a Germany whose actions he did not approve of, and Ehrenfest now realized he might never again see Einstein, the man with whom he had shared some of the happiest moments of his life.

Nor were teaching and research the solaces they had once been. Ehrenfest felt neither the energy nor the inclination to begin learning the details of the emerging field of nuclear physics. In the fall of 1932 he published "Some Exploratory Questions Regarding Quantum Mechanics," laying out in a journal a number of the issues that continued to trouble him. Trying to remain intellectually active, he turned his fine critical mind to understanding some earlier concepts he felt had not been sufficiently clarified.

As soon as Pauli saw Ehrenfest's observations, he wrote him a long

Portrait of Ehrenfest/Faust

letter, first letting Ehrenfest know how delighted he was to see the questions he posed and then going on to describe how these were the very same ones that had bothered him while preparing his recent *Handbuch* work on quantum mechanics. Pauli went on to formulate answers for them as best he could. Ehrenfest was thrilled by the letter and wrote back that he had hesitated to submit the paper for more than a year because he was afraid that Bohr and Pauli would think less of him, viewing the effort as without merit. As Ehrenfest/Faust says in the skit,

> All *doubts assail me; so does* every *scruple;*
> *And Pauli as the Devil himself I fear.*

Ehrenfest's obvious pleasure at hearing from Pauli was only a momentary reprieve from his darkening mood. Increasingly obsessed by his losses, by family concerns, and by what he regarded as his own shortcomings, Ehrenfest began to contemplate suicide. He had already written in August 1932 to a half-dozen friends, including Bohr and Einstein, expressing this wish, but he never mailed the letters. In 1933, as matters worsened, the plan to end his life began to seem irrevocable to him.

Twenty years earlier Lorentz had worried about the future of physics in his beloved Leiden, accepting retirement only when certain that he was leaving it in good hands. Ehrenfest felt he must do the same. Casimir, his former student, the one he had brought to the first Copenhagen meeting, was now in Zurich, having succeeded Peierls as Pauli's assistant. Casimir was hoping to stay there at least one more year, but at Easter 1933, he received a puzzling letter from Ehrenfest, asking him to come back to Leiden in September. In its final sentence, using his affectionate nickname for Casimir, Ehrenfest pleaded, "O Caasje, put your broad shoulders under the wagon of Leiden physics." Casimir and Pauli did not understand why Ehrenfest was making this request or the strange language he was using, but they quickly agreed to honor his wishes.

Easter holidays would normally have been the time for the Copenhagen meeting, but because Bohr was away on a prolonged visit to the United States, schedules were rearranged so that it could be held at the end of the summer, with a starting date set for Sunday, September 13.

The meeting's front row is shown in photograph 14: Bohr, Dirac, Heisenberg, Ehrenfest, Delbrück, and Meitner. It was a formidable group, whose prominent place in the seating, symbolic of their role in physics, was not by chance. As the meeting's participants filed into the lecture hall, there was no need for a "reserved" sign on those front-row seats. Their occupants were respected and revered. There was no doubt who would occupy them or why they were there.

Pauli normally would have been placed in that front row as well. But he missed the meeting again, this time because, exhausted from his review of quantum mechanics in the *Handbuch*, he decided to take a September vacation in southern France. The sessions were very stimulating even without him. Returning to Leipzig, Heisenberg wrote Bohr that it had been a long while since he had learned as much during a short span as he had during the Copenhagen week. Pauli knew it was a pity that he had not participated, but he thought he would catch up with his Copenhagen friends in Brussels at the Solvay conference, to be held in October. He certainly never suspected that he would be missing his last chance to see Ehrenfest.

After the meeting ended on September 20, Ehrenfest went straight back to Holland. On the twenty-fifth he traveled from Leiden to Amsterdam to visit his youngest son, fifteen-year-old Vassilji (known as Vassik), under care in an institution for severely retarded children. In a horribly misguided notion that his son should be put out of his misery at the same time as he, Ehrenfest took him to a nearby park, drew a revolver from his pocket, first shot Vassik, and then turned the gun on himself. Vassik was blinded by the bullet, but did not die. His father's death was instant. That Copenhagen front-row photograph may have been the last one ever taken of the Leiden physicist.

Three days after Ehrenfest's death, Dirac wrote Bohr describing to him how guilty he was feeling for not having done anything to prevent the suicide. He told him that on running into Ehrenfest at the meeting's end, he had thanked him for his valuable contributions to the discussions, only to have Ehrenfest become agitated, break down in tears, and then address him haltingly, with a single sentence uttered before leaving abruptly: "What you have just said, coming from a young man like you, means very much to me, because, maybe, a man such as I feels he has no longer the force to live."

We will never know whether Ehrenfest let his guard down in a way he would not have done with Bohr and other old friends or whether Dirac, who always took things to mean exactly what the words said, heard something that others discounted. Bohr reassured Dirac that in any case his intervention would not have made a difference.

In light of the suicide, it seems almost ghoulish to go back to the 1932 *Faust* parody and read into it further parallels flowing from the association of Doctor Faust with Ehrenfest. The murder of a child occurs in Goethe's play, though it is Gretchen, not Faust, who kills the fruit of their love. Blindness is inflicted, though on Faust himself at the end of *Part II.* But the main identification of Faust and Ehrenfest is one of personalities. The twenty-five-year-olds knew Ehrenfest was overly self-critical, but did not understand the underlying despair.

The opening monologue of *Faust,*

> I have, alas! Philosophy,
> Medicine, Jurisprudence too,
> . . . .
> Then I have neither goods nor treasure,
> No worldly honor, rank, or pleasure;
> No dog in such fashion would longer live!
>
> (*Faust, Part I,* 1–2, 21–23)

appears in the skit in the guise of Ehrenfest/Faust reciting all the subjects a quantum physicist needs to learn. He concludes by saying,

> *. . . nor hound could bear my lot,*
> *So I'm the Critic, sad and misbegot.*

The reader should not remember Ehrenfest as "sad and misbegot." As Pauli observed in his obituary of him, he was "a man of scintillating intellect and wit, intervening in discussions with mordant criticism, and at the same time with profound insight into the scientific attitude, drawing attention to some essential point hitherto unnoticed or insufficiently regarded."

Less than a month after Ehrenfest's death, a saddened group gathered in Brussels for the seventh Solvay conference. Even though its participants were deeply worried about the rise of Hitler and could not help noticing the absences, such as Einstein's, that had resulted, the conference was held to be a great success from the point of view of physics. But the attendees missed the great undercurrent provided to the previous two Solvay conferences by the debate between Bohr and Einstein about the meaning of quantum mechanics. They also mourned the short man with the round glasses and the lively smile they remembered walking so often with the two of them.

The seventh conference's topic, as always the central subject of physics at the time, was "The Atomic Nucleus." During the week in Brussels, Chadwick reviewed the discovery of the neutron and its implications for nuclear physics; Heisenberg delivered the major address on the subject of nuclear forces; Blackett, Meitner, and others discussed the experiments proving that positrons exist; and Dirac offered his new thoughts on electron theory. Pauli presented the neutrino hypothesis in public for the first time, maintaining once again his firm conviction that energy was conserved even in nuclear reactions. The event was brimming with new ideas, ushering in the era of nuclear physics for all its good and all its evil.

Fermi, who had been fairly quiet during the conference, went back to Italy convinced that physics' immediate future lay in nuclear research. Within weeks he had formulated his theory of the weak interactions and a few months later began bombarding atomic targets with neutrons. Since Fermi was not as politically astute or as visionary as Szilard, he may not have thought at the time of chain reactions, much less of nuclear weapons. But within the next ten years they were his central concern. He would immigrate to the United States (his wife was Jewish), lead the effort to build the first controlled nuclear chain reaction, and become one of the key physicists in building the atomic bomb.

By 1934 the days of carefree exchanges in Copenhagen had come to an end. Physicists' Blegdamsvej conversations were now sprinkled with questions about which of them had left Italy or Germany or the Soviet Union, who was staying, what were conditions really like back in their home country, where could one find a safe haven, and what might the future bring. Gamow had been the first of their crowd to be denied governmental permission for foreign travel, but restrictions of one sort or another were now cropping up for all of them. Still in the background, but swiftly approaching, was the day when their beloved physics would provide the most powerful weapon humans had ever known.

Quantum theory and the submicroscopic world of the atom would become major factors in the power games of nations. The realities of nuclear chain reactions would change humanity's worldview. Physicists' discussions would shift from harmless parodies of *Faust* to the making of ominous Faustian bargains. Heisenberg and Weizsäcker continued to remain close, except for a brief period when Heisenberg courted and was then rejected by Carl Friedrich's younger sister Adelheid. Later the two physicists became critical members of the German nuclear weapons team. Oppenheimer led the United States group, aided by many of the physicists who have figured in this story: Bethe, Chadwick, Fermi, Frisch, Peierls, Segrè, and Weisskopf all went to Los Alamos. Bohr made a dramatic escape from Denmark and eventually joined them in the New Mexico desert, acting as a guiding conscience and elder statesman more than as a solver of technical questions.

Though they did not know it then, the 1932 Copenhagen meeting was the last moment in the lives of these young scientists when they could still laugh together at a struggle between the Lord and Mephistopheles. The relative innocence of those days was soon lost. Yet, remarkably, the Copenhagen spirit of free and unyielding inquiry has survived, a tribute to its deep roots. Meetings are now larger, experiments bigger, and research more specialized. The long letters of yore have been replaced by e-mails destined for oblivion, new results are transmitted instantly via the Internet, and computational methods have progressed to an almost unimaginable degree. But you can still go into physics labs anywhere in the world and find two or three theorists arguing in front of a blackboard. They are laughing at the same type of jokes that amused the Blegdamsvej *Knaben*, still putting on skits that poke fun at their elders, still being appropriately critical and disrespectful of their elders, and still looking for the big ideas that will unlock the secrets of nature.

Epilogue: Or What Happened Afterward to the Front Row's Other Six

Ye wavering shapes, again ye do enfold me,
As erst upon my troubled sight ye stole;
Shall I this time attempt to clasp, to hold ye?

Goethe, *Faust, Part I,* Dedication, 1–3

How Meitner Discovered Nuclear Fission

MANY OF GERMANY'S leading physicists left soon after Hitler's takeover and the subsequent passage of racial laws. This was easier for the young ones, more inclined to start over elsewhere, but older ones like Max Born, Otto Stern, and of course Albert Einstein departed then as well. Others did not. Although Lise Meitner's Jewish background made her vulnerable to the galloping German anti-Semitism, she hesitated, not wanting to abandon her thriving laboratory or alter the life she had struggled to establish. She also felt protected by her Austrian citizenship, but after the Anschluss (annexation) of Austria by Germany in March 1938, that protection ceased to exist. By June 1938 it was clear she was about to be dismissed from her institute. Furthermore her old passport was now invalid, and her application for a new one, as a German citizen, was rejected. In effect, she and other Jews were now finding the borders sealed.

On July 13, without a passport or any of her possessions, Meitner surreptitiously entered the Netherlands at a small, lightly manned border crossing. Soon after that, she flew to Copenhagen and then, with

Bohr's assistance, took refuge in Sweden. Five months later, during Christmas vacation, her physicist nephew Otto Frisch, working at the time in Copenhagen, came to spend a few days with his favorite aunt. He found her pondering some curious nuclear physics results Otto Hahn and his collaborator Fritz Strassmann had very recently obtained from experiments that, until a few months earlier, Meitner had also worked on. Hahn could then have walked down the hall and asked her what his findings meant, but he now had to do it by letter.

Hahn and Strassmann had been bombarding uranium nuclei with neutrons, expecting to obtain radium nuclei, slightly lighter than uranium nuclei, as a by-product of the decay that followed neutron absorption. Hahn was, however, coming to a strange conclusion; he wrote Meitner, "There is something about the radium isotopes that is so remarkable that for now we are telling only you." The radium looked more like barium, seemingly impossible because barium, a much lighter element, was far lower on the periodic table than radium.

On the morning of Christmas Eve, 1938, Frisch and Meitner went for a walk. During the walk they realized the data could be explained if absorption of a neutron caused a uranium nucleus to split into two smaller remnants, one of which was a barium nucleus. This would be a true nuclear breakup and the subsequent release of energy enormous even by nuclear physics standards. A few days later Frisch went back to Copenhagen and spoke to Bohr, who appreciated their idea immediately. Frisch remembers Bohr saying, "Oh, what idiots we have all been! Oh, but this is wonderful! This is just as it must be!"

Frisch soon verified in his Copenhagen lab the main feature of their conjecture by making a direct observation of the nucleus dividing. Groping for a name, he decided to call the process nuclear fission. The energy realized in the uranium nucleus's breakup was more than a billion times greater than that carried by the neutron that induced the fission. In addition, two or more neutrons were produced as the division occurred. This meant that the two key factors in the production of a chain-reaction nuclear weapon had now been realized, at least in principle.

The technical details of building a bomb still seemed overwhelming because ordinary uranium would not undergo fission. Two candidates were available as material for an explosive device, each of them requiring a gigantic industrial effort if enough was to be accumulated for a sizable weapon. One could either produce a sizable amount of the transuranic element plutonium or separate a large quantity of the rare uranium isotope U-235 from ordinary uranium ore. Both approaches were undertaken in the United States during the early 1940s, and both eventually resulted in bomb production. The first type of bomb was used on Nagasaki and the second type on Hiroshima.

Meanwhile Meitner, isolated in Sweden during the war, had no idea of the progress toward nuclear weapons, although some in the popular press after Hiroshima portrayed her as fleeing Germany in 1938 with the secret of the bomb and then turning it over to the Allies. She remained in Sweden until 1960, refusing postwar offers to return to Germany. Moving to Cambridge, England, where Otto Frisch was then a professor, she died there peacefully in 1968. Otto Hahn, who had died a few months earlier, was the recipient of the 1944 Nobel Prize in Chemistry "for his discovery of the fission of heavy atomic nuclei."

How Bohr Lived Happily Ever After

BOHR CONTINUED TO BE what he had always been, a deep thinker, a warm and caring human being, a model of integrity, and a father figure to generations of scientists.

In the summer of 1932, just months after the meeting at which the "Copenhagen Faust" had been performed, the Bohr family moved to Aeresbolig, the magnificent nineteenth-century Copenhagen villa willed to the Danish government by the Carlsberg Brewery's founder with the proviso that it be used by the most prominent Dane in the field of science or of the arts. The Bohrs' first guests in Aeresbolig, arriving for a ten-day stay in September 1932, were chosen with particular care: Lord and Lady Rutherford.

But Bohr's favorite residence was Lynghuset, a country dwelling in Tisvilde, thirty miles from Copenhagen along the Sjælland coast. Feeling strongly the need for a place of their own, the Bohrs had purchased it in 1924, when they were still living on the upper floors of the institute building on Blegdamsvej.

Lynghuset was set on a large sand dune at the edge of a beech and pine forest, the trees having been planted two hundred years earlier to anchor the shifting sands. The thatched-roof house was simple but comfortable, an ideal retreat for the growing family. Bohr went there to relax, but he would almost always bring along one or two young physicists because talking with them was a form of relaxation for him. They would work together in the small study fifty yards from the house, alternating physics discussions with sailing or playing games with the Bohr boys.

But it seems that some tragedy falls on even the happiest of lives. For Bohr the worst came on the second of July, 1934. He was sailing off the coast with a few friends and his oldest son, Christian. A sudden wave swept across the boat, hurling the teenager into the water. They threw him a life buoy; he almost reached it, but then was dragged under. Bohr had to be restrained by his friends to prevent him from jumping into the waters.

A visit to Tisvilde by Pauli was a special treat for Bohr, his favorite critic coming to see him in his favorite place. Proposing such a stay in August 1929, Pauli wrote, "The temptation is great to visit you then around the middle of August in your house without water pipes—I would bring plenty of candles along," to which Bohr quickly replied: "How nice it would be . . . I am longing to discuss many questions with you . . . You need not bring along candles; we have now got electric light out here. I hope you are going to stay for a long time."

During the 1930s Bohr worked ceaselessly to provide refuge for scientists escaping persecution. When World War II began, he redoubled his efforts, leaving Denmark only in extremis when it became clear he would be deported (Bohr's mother was Jewish). Alerted by Swedish diplomats that a roundup of Jews was imminent, he, Margrethe, and a few friends made an illegal nighttime crossing to Sweden by fishing boat on September 29, 1943. Bohr immediately proceeded to Stockholm, where he secretly made a direct, personal appeal to the king, asking him to proclaim that Sweden would give asylum to Danish Jews. The appeal was successful, and an announcement was made on October 2. Over the next two months more than 90 percent of Denmark's Jews, including all of Bohr's family, were rescued by being hidden and then transported via fishing boats out of Danish territorial waters to Swedish vessels that took them to their new refuge.

Meanwhile there was concern that German agents in Stockholm would assassinate Bohr. To protect against this possibility, he was flown to Britain on October 5 in a light bomber plane able to fly high enough to avoid the German-operated antiaircraft guns in occupied Norway. The bomb bay had been specially converted to house a passenger for the flight, but Bohr, failing to turn on the oxygen properly, soon became unconscious. The pilot, aware of what was happening, descended to lower altitudes as soon as possible, and Bohr survived the trip without suffering any harm. A few months later, in the early winter of 1943, he flew to the United States, traveling to Los Alamos, where he began encouraging physicists to think about how to achieve disarmament and world peace after the war was over. These were issues that concerned him for the rest of his life.

That life was brought to an end by a heart attack on a Sunday afternoon in 1962 as he was taking a nap in his Aeresbolig bedroom. He and Margrethe had celebrated their fiftieth wedding anniversary a few months earlier.

How Dirac Got Married

DIRAC'S LIFE CONTINUED on an even plane, apparently free of the ups and downs that characterized the lives of so many of his contemporaries. The Nobel Prize, Lucasian professorship, and the many other honors that came to him did not change his steady pace. Despite the port, fine wines, and cigars at Cambridge college high tables, Dirac continued leading the ascetic life that suited his tastes.

The idea of the solitary Dirac ever having a relationship with a woman, much less marrying one, seemed strange to most who knew him, but it happened. In 1934 he left Cambridge to spend some time in Princeton. There he became reacquainted with Eugene Wigner, a prominent theoretical physicist and a professor at Princeton University. They had met in Göttingen almost a decade earlier.

While Dirac was there, Wigner received a visit from his sister Margit. Recently divorced, with two children, she lived in Budapest. Wigner's sister and Dirac became friends, and the next summer he visited her in Budapest. They were married in 1937. Dirac then moved out of St. John's College to join his new family, soon increased in size by two children the Diracs had together, in a Cambridge house. In 1970 they relocated to Tallahassee, Florida, where Dirac took up a professorship at Florida State University after retiring from Cambridge.

Though he lived until 1984 and continued to be active, none of the research that he did in the last half century of his life matched in brilliance or importance the work that he had done in the period between 1925 and 1934. He recognized this and accepted the fact, saying simply, "My own contributions since the early days have been of minor importance."

People were sometimes puzzled when Dirac introduced his wife to them as Wigner's sister, but he was only being logical. That is how he had come to know her.

How Heisenberg Inspired His Friend to Paint Like Titian

HEISENBERG ALSO MARRIED IN 1937, to Elisabeth Schumacher. Exactly nine months later she gave birth to twins, a boy and a girl, leading Pauli, with an obvious reference to electrons and positrons, to congratulate them on their successful "pair production."

The second half of the 1930s was a difficult time for Heisenberg.

Some accused him of being a so-called White Jew for refusing to condemn the likes of Born and Einstein; others called him a Nazi for his continued tacit support of the regime. Matters became even worse when World War II started and he assumed a position of leadership in the German nuclear weapons program, aided once more by Weizsäcker. His aims in that project have remained unclear, leaving it still a hotly debated topic, again with two sides. Some argue that he did all he could to build a German atom bomb while others maintain he deliberately sabotaged the program, with many left believing Heisenberg never resolved his own ambivalence. Nor did his subsequent presentations ever settle the question decisively.

After the war was over, he returned to a position of prominence in German science as director of the Kaiser Wilhelm Institute for Physics in Göttingen; it was subsequently renamed the Max Planck Institute for Physics. Though Heisenberg's papers on nuclear forces in 1933 had been very significant advances, his influence on the intellectual evolution of his subject waned after that. The curse of the over-thirties in theoretical physics had struck, though unlike Dirac, Heisenberg refused to accept its verdict.

He persisted in thinking he could change the course of physics, in particular through the development of his so-called nonlinear theory. In 1957 he consulted with Pauli on its potential. Though their exchanges had continued, they had not actually collaborated on any research since 1930. At first Pauli was enthusiastic about joining forces with the man he had known for almost forty years, thinking it might be like old times for the two of them. They began working together, and a preliminary version of a manuscript was quickly prepared.

Delving deeper into the problem and listening to the comments of Bohr and others, Pauli's critical faculties took over and he withdrew from the project. Meanwhile Heisenberg, refusing to be discouraged, announced the discovery he thought he had made to an audience in Göttingen and later to a crowd of almost two thousand gathered to celebrate the hundredth anniversary of Max Planck's birth.

During the summer of 1958, Heisenberg spoke at a session entitled "Fundamental Ideas" of an international conference in Geneva. Introducing him, Pauli, the chairman of the session, began by saying that contrary to the session's label, he didn't think Heisenberg would present any fundamental ideas in his talk. Later mocking Heisenberg's notion that the theory only lacked technical details, Pauli sent friends a drawing of an empty rectangle, with a comment underneath: "This is to show that I can paint like Titian; only technical details are missing. —W. Pauli." As usual Heisenberg was the more daring of the two, but in this case the more critical and the more sarcastic Pauli was right.

In September 1958 Heisenberg transferred his Max Planck Institute to his beloved Munich, where more than twenty years earlier he had hoped to succeed Sommerfeld as professor. Pauli's death in December 1958 prevented the two from healing their breach, as they almost certainly would have, since it was not the first time Heisenberg had been subject to Pauli's barbs.

Heisenberg died of cancer at his Munich home in 1976.

How Pauli's Anima Made Him Leave the United States

THE ANALYSIS by Jung and his assistant, or perhaps just the passing of time, seems to have healed Pauli of what he had called his neurosis. In 1933 common friends introduced him to Franca Bertram, a woman as strong-willed as he was. The two of them, getting along very well from the start, were married in 1934. This time the marriage was a happy and a stable one.

As World War II approached, Pauli realized his Jewish roots put him at risk. Since he had not yet obtained Swiss citizenship, Germany could, at least in principle, reclaim him. In the late summer of 1940 the Paulis fled to the United States, taking up residence in Princeton. When he received the 1945 Nobel Prize in Physics, the Institute for Advanced Study, which had just offered him a professorship, held a party in his honor. On that occasion Einstein, unannounced, gave an impromptu address that moved Pauli deeply: "I will never forget the speech about me, and for me, that he gave at Princeton in 1945 after I got the Nobel Prize. It was like the abdication of a king, installing me as a kind of elected son, as his successor."

Though Pauli felt very strongly what an honor and privilege it was to be a colleague of Einstein's and was tempted to remain in the United States, he also knew he needed a politically neutral environment. Writing to a psychologist friend, Pauli expressed the unease he was experiencing at this time and his desire to return to Switzerland: "As in Austria during the First World War, in this year (1945) I suddenly had the feeling that I was placed in a 'criminal' atmosphere—and this at the time those A-bombs were dropped . . . My anima became very irritable and occasionally produced eruptions of anger until I had departed the U.S.A."

Pauli returned to Zurich, where he once again led a productive group of theoretical physicists including, as had been the case earlier, a large number of visitors from distant shores. But he too was never as influential as he had been earlier.

He died in Zurich in December 1958 of a cancer that was not diagnosed until days before his death. His last writing in physics was a piece entitled "On the Earlier and More Recent History of the Neutrino," an article whose content ranged from a description of the early work on the subject to the recent announcement by Clyde Cowan and Frederick Reines that they had succeeded in the direct observation of neutrinos using a nuclear reactor. Greatly pleased by this milestone, he sent Lise Meitner a copy of the article, letting her know that it was meant as a small present for her eightieth birthday.

After World War II, Pauli gradually acquired a new and less portentous nickname than Scourge of God. The old caustic personality remained, but Pauli's steady refusal to contribute to political, much less military-related, causes, his unwavering interest in intellectual matters, and his insistence on supporting only work of the highest quality, no matter who had done it, earned him a new sobriquet, *The Conscience of Physics.* It was the same title that been given decades earlier to his old friend Paul Ehrenfest.

How Delbrück Became a Biologist

WHILE WORKING AS the "family-theorist" in Meitner's laboratory, Delbrück often returned to see Bohr and the continually changing group at Blegdamsvej. After the April 1932 meeting he had gone back to Berlin but in August was once again in Copenhagen. Sleepily stepping out of the carriage of the overnight train from Berlin, Delbrück was surprised to see Bohr's trusted collaborator Leon Rosenfeld waiting for him with a message from Bohr. Delbrück was to go straightaway to the great meeting hall of the Rigsdag, the Danish parliament, where Bohr was about to deliver the opening lecture at the International Congress of Light Therapists. The talk, entitled "Light and Life," was largely a ceremonial affair, held in the presence of the Danish crown prince, the prime minister, and other dignitaries. Knowing Delbrück's interest in biology, Bohr wanted him to attend. This turned out to be the talk that changed Delbrück's life.

In 1932 some biologists still clung to the notion that special forces not described by the ordinary laws of physics and chemistry were responsible for the existence of life. Most, particularly those trained as biochemists or physiologists, objected to this belief. Bohr stepped into the breach that year, trying to suggest a way of reconciling their two very different views of life, just as he had once reconciled the picture of light as particles and waves. He now asked: Could complementarity apply to life itself? Perhaps the distinction between living and nonliving was not so easy to understand. Might the act of measurement be a critical step in the assessment?

The issue was whether one could simultaneously observe life and the mechanisms responsible for it. Even though living matter is composed of the same atoms as inanimate matter and obeys the same laws of physics and chemistry, Bohr thought there might be some intangible quality that differentiates living from nonliving, a quality that could not be precisely quantified. He was not alluding to a divine spark, but rather to something analogous to the impossibility of saying with certainty whether light is a photon or a wave. Bohr's conjecture was provocative, as it was meant to be, but in the end it turned out to be wrong. DNA and RNA are the answer to life, not complementarity.

But the potential connection between physics and biology had surfaced. Would Delbrück have become a biologist if he had not thought he might find something like complementarity at the root of life's existence? Would molecular biology have developed as it did without the structure imposed on it by the discipline Delbrück had learned from Bohr? Many who know the field think the answer to both questions is no. When James Watson wrote Delbrück in 1953 that he and Francis Crick had found their double helix model of DNA, Delbrück was struck by the elegance of the structure, but disappointed by its simplicity. Hearing that life did not require any basic new principles, he remarked that it felt to him as if the hydrogen atom had been fully explained in the 1920s without the need for quantum mechanics.

By then Delbrück had become a Bohr-like figure in the new field of molecular biology. He had created at Cold Spring Harbor Laboratory on Long Island and at the California Institute of Technology Copenhagen-like atmospheres for young biologists. He did the same in Germany, returning there for two years in the early 1960s to help start a modern molecular genetics research group at the University of Cologne. It was inaugurated June 22, 1962, in a ceremony at which Niels Bohr delivered the principal address. Entitled "Light and Life Revisited," it was the last formal speech Bohr delivered—the version of it he was preparing for publication was unfinished at his death.

Delbrück may have been disappointed that no analogue of complementarity had been necessary to explain the existence of life, but he, together with Salvador Luria and Alfred Hershey, with whom he shared the 1969 Nobel Prize in Physiology or Medicine, had shown how bacteriophages, the viruses that attack bacteria, were the simplest and most efficient tool for studying genetics. Such a virus, an entity made of nucleic acids surrounded by a protein shell, can attack a bacterium and exchange genetic material by a set of rules Delbrück and Luria worked out. In many ways the importance of these findings was comparable to Bohr's 1913 discovery of the rules for the hydrogen atom's radiation.

Moreover, like Bohr's impact on physics, Delbrück's personality had a great importance for the new subject of molecular biology. As Watson wrote, "Max's personal integrity and abhorrence of pomposity, together with his intellectual honesty and openness, left us no doubt as to how we were to behave if we were to be worthy of the scientific life."

Perhaps these qualities are the basis of the deep lifelong link between Pauli and Delbrück. Luria, in his memoirs, remembered his first meeting with Delbrück. At the time both were recent refugees from Europe. It was New York City in late December 1941, only a few weeks after Pearl Harbor:

> Max took me to dinner with two other scientists, one of them the great physicist Wolfgang Pauli. I was properly intimidated, but Pauli simply asked me "*Sprechen Sie Deutsch?*" and without waiting for a reply proceeded to eat and speak German so prodigiously fast that I understood not a word. I would have been even more scared had I known of Pauli's classic remark "So young and he has already contributed so little."

Fortunately young Luria soon contributed a great deal.

As for Pauli, his last significant letter, written two months before he died, was a twelve-page one to Max. Beginning with a discussion of the festivities for Lise Meitner's eightieth birthday, it ends on a more personal note: "I cannot forget the dear manner, which goes back to old times, when you took leave of me. I have the impression that here something has been renewed between us and that is important."

Delbrück would hike and camp with young biologists in the Southern California desert rather than walk with them along the North Sea shore of Denmark, but the spirit was the same as Bohr's had been. On the other hand, rather than Bohr's polite, "I don't mean to criticize, only to understand," Max remained famous for his rudeness, for walking out of seminars if the speaker didn't get to the point. If he stayed to the end, he was known to then say, "This was the worst seminar I ever heard." If a colleague presented him with a conclusion, he would often comment quickly, "I don't believe a word of it," urging the colleague to try harder to convince him. So we see that though the spirit was Bohr's, the style was Pauli's. He sometimes quipped to his friends that he wanted to be "God and Mephisto, all in one."

Delbrück died in 1981 in California.

Coda

THIS BOOK BEGAN with the story of Katia and Kate, my mother and my mother-in-law, arriving in Munich in the fall of 1918, and I don't want to leave it without telling you what happened to them in their later life.

Kate remarried after the Nazis murdered her first husband. Her second tie was to a physicist and famous mountaineer named Hermann Hoerlin. She came with him to the United States and first lived in Binghamton, New York, where their daughter Bettina, now my wife, was born. Moving later to Los Alamos, they befriended many of this story's protagonists. Kate's last years were spent near Boston, close to her two Munich-born daughters. She died there in 1985 and was buried in Mount Auburn Cemetery.

Katia married Angelo Segrè and lived with him in Florence. They moved with their two sons to the United States just before World War II began, but shortly after the war ended, they went back to Florence. When my father died many years later, Katia decided to return to the United States in order to be close to her sons and grandchildren, who by then were living there. She moved to Boston, where my older brother, her Munich-born son, resided. When she died in 1987, we decided to bury her in the Mount Auburn Cemetery and went there to inquire about the availability of plots. They recommended a spot to us on a little hillside that overlooks a beautiful pond. When I went to examine the site together with Bettina, Kate's daughter, we discovered to our amazement that Katia's intended plot was eerily close to Kate's. Their paths had crossed once again.

Good-bye Katia, Kate, and all you *wavering shapes* out of the past. We thank you.

Acknowledgments

WRITING THIS BOOK has been a labor of love, allowing me to spend time in the company of many of the intellectual heroes of my youth. During the writing, I have gained information and inspiration from historical accounts, from memoirs written by those who participated in the events, and, even more vividly, from the letters these participants wrote to one another as they advanced toward their discoveries. Though modern technology has provided us with many advantages, I fear there will be nothing quite like the record provided by such correspondence.

There are two individuals whose contributions I would like to single out, albeit for very different reasons. The first is the late Abraham Pais, a distinguished theoretical physicist, one whose work I had always admired. During his career, he came to know personally most of the main characters of this story, first during his stay as a postdoctoral fellow in Copenhagen and later while a professor at the Princeton Institute for Advanced Study. With an interest in the history of his profession, he realized in later life that he was one of the few remaining individuals to have had this experience. Recording what he knew and what he came to learn, Pais eventually produced, inter alia, biographies of Bohr and Einstein, and a history of twentieth-century physics' quest for studying the ever smaller components of matter. His work has been invaluable for me.

The second author I wish to single out, the late George Gamow, was also an extraordinarily distingished physicist, as I hope is clear from the text. In addition to his own pioneering research, he devoted considerable time to the writing of books for the general public, introducing readers from around the world to the wonders of modern physics. An inveterate practical joker and humorist as well, it is unlikely that the tradition of skits at the yearly Copenhagen meetings would have been set in motion without him. I am grateful to him and to his wife, Barbara, for their translation and publication of the *Blegdamsvej Faust* that plays such a role in this book. I am also deeply indebted to Gamow's son Rustem Igor Gamow and to his wife, Elfriede, for graciously allowing me to quote from the senior Gamows' work and make use of the illustrations that George Gamow drew. Their presence certainly enriches this book.

I asked four colleagues in the physics community to read through the manuscript, rooting out errors. Ralph Amado, Michael Cohen, Kurt Gottfried, and Robert Harris generously consented. Their advice was invaluable as were conversations about the book with Phil Nelson. Any remaining mistakes are of course purely my own, but there would have been more had it not been for these men.

Earle Spamer at the American Philosophical Society assisted me in consulting the society's fascinating collection of archival material in the history of quantum mechanics. Heather Lindsay at the American Physical Society and Felicity Pors at the Niels Bohr Archives in Copenhagen were also very helpful to me. In addition I would like to thank my home institution, the University of Pennsylvania, for granting me a sabbatical in which to pursue the writing of the book, and the Liguria Foundation for welcoming my wife and me for a marvelous stay at its Bogliasco Center.

I am grateful to my agents, John Brockman and Katinka Matson, for their creative approach to writing about science and for encouraging me to pursue my own interests along this direction. Wendy Wolf at Viking, assisted by Clifford Corcoran, has been a wonderful editor. She helped reshape the manuscript, gave it at times a focus it lacked, and always provided sage advice. At one point, after presenting her with what I thought was an essentially complete version, she gently said that it was not too bad, but why didn't I try to write a really good book. I hope I have succeeded, but in any case the book that appears is surely better because of her prodding and her criticisms. I also wish to acknowledge Don Homolka's contributions as a copy editor.

I would like to remember at this point three individuals whose spirit helped steer me. The first is my father, Angelo Segrè, my initial guide to the world of learning. The second is Viki Weisskopf, who I came to know as a student early in my career and as a brother-in-law many years later. His carrying with him the spirit of Bohr and Copenhagen was always an inspiration. The final individual of the three is Bram Pais, whose scholarship remains a touchstone for all of us in the world of physics.

As with my previous book, my final thanks go to my wife, Bettina Hoerlin, for her love, her patience, her good cheer, and, not least, for her help and criticism. At a certain point in our sessions Wendy Wolf would invariably say to me, "What does Bettina think of this?" When I would give her my reply, Wendy would smile and then say, "You better listen to her." So the final word in the acknowledgments is rightly to say that once again, this book, with love, is for Bettina.

Notes

To distinguish different works by the same author, the publication date of the reference is given, for example, Dirac (1926) and Dirac (1929). The publication year is otherwise omitted.

Abbreviations

AHQP	Archives for the History of Quantum Physics. On deposit at the American Philosophical Society, Philadelphia, and the American Institute of Physics, College Park, Maryland.
CW	Niels Bohr. 1972– . *Collected Works.* 11 vols. Amsterdam: North-Holland.
NBA	Niels Bohr Archives. Niels Bohr Institute, Copenhagen, Denmark.
PSC	Wolfgang Pauli. 1975. *Scientific Correspondence with Bohr, Einstein, Heisenberg, and Others.* 6 vols. Ed. K. v. Meyenn with the assistance of A. Hermann and V. Weisskopf. New York: Springer.

Introduction

The connection between Bohr and Einstein is discussed particularly sensitively by the late Abraham Pais, who knew them both well and has written excellent and lengthy biographies of each of them.

3 *Bohr's influence . . . that of Albert Einstein:* Heisenberg (1986), vol. 4C, p. 144, also quoted in Pais (1991), p. 14.

4 *his opinions . . . holds the whole defining truth:* Letter from Einstein to Bill Becker, March 20, 1954, in Calaprice, p. 73.

5 *Pauli possessed . . . his true thought directly:* Weisskopf, p. 84.

5 *I met Pauli . . . into atomic physics:* Letter from Bohr to Bartel van der Waerden, July(?) 1959, in CW, vol. 5, p. 507.

6 *When more than thirty . . . one is indeed:* Goethe (1905). All quotations from *Faust* are from this translation, by Anna Swanwick.

7 *the United States into one huge factory:* Blumberg and Owens, p. 89, also quoted in Rhodes, p. 294.

Chapter 1: Munich Then and Now

10 *Pauli's character . . . just too much:* Werner Heisenberg in *International Atomic Energy Agency Bulletin,* Special Supplement, 1968, p. 45, also quoted in Meyenn and Schucking, p. 43.

11 *Yes, Herr Geheimrat . . . perhaps I would prefer:* Weisskopf, p. 85.

12 *No one studying . . . sureness of critical appraisal:* Einstein (1922), also quoted in Meyenn and Schucking, p. 44.

16 *I have seen dictatorship . . . it just couldn't happen:* Blumberg and Owens, p. 51.

Chapter 2: The Changing Times

The 1920s

18 *museums are cemeteries . . . Victory of Samothrace:* Marinetti (author's translation).

The Birth of the Quantum

23 *It was an act of desperation . . . two laws of thermodynamics:* Hermann, p. 74.

25 *fly in the cathedral . . . a gnat in Albert Hall:* Cathcart, p. 6.

Why Copenhagen?

25 *To me, . . . a second father:* CW, vol. 1, pp. 106 and 533, also quoted in Pais (1991), p. 129.

28 *I learned optimism . . . physics from Bohr:* Heisenberg (1986), part C, vol. 1, p. 4; also quoted in Pais (1991), p. 163.

The Meetings Begin

30 *In this situation . . . more or less fortuitous:* Heisenberg (1925), p. 33, translated in Van der Waerden, p. 262.

32 *Now you are going . . . the life of a young physicist:* Casimir, p. 89.

32 *I am bringing . . . still needs thrashing:* Casimir, p. 91.

33 *Yesterday evening . . . everybody was clearly shaken:* Segrè (1993), p. 122.

Chapter 3: Goethe and Faust

The literature on Goethe is of course vast. For an easily readable introductory essay, with a guide to further readings, I recommend Daniel Boorstin's *The Creators*, chapter 61, pp. 598–613. My quotations from *Faust* are from a nineteenth-century translation (Goethe [1905]), which may represent more faithfully how the Copenhagen physicists knew the great drama. I have, however, given some references to translations more attuned to the modern ear.

All quotations from the "Copenhagen Faust," sometimes referred to as the "Blegdamsvej Faust" after the street on which Bohr's institute was located, are from the English translation that George and Barbara Gamow prepared. It appears on pp. 165–214 of the Dover 1985 reprint edition of Gamow's *Thirty Years That Shook Physics* (Gamow [1966]). The only exceptions are the lines that appear in the section of chapter 5 entitled "Old Age Is a Cold Fever." These were apparently omitted in Gamow's translation. They can be seen, in the original German, in Meyenn, Stolzenburg, and Sexl, p. 308. The translation of these lines is by the present author.

In the Glow of Goethe

42 *For Goethe . . . accessible to our senses:* Heisenberg (1971), p. 123.

The "Copenhagen Faust"

46 *I would telephone . . . arrange matters:* Gamow (1970), p. 111.

47 *The Blegdamsvej* Faust *. . . certain parts of the play:* Gamow (1966), pp. 168–69.

48 *Professor Bohr . . . frequently quoted from Goethe:* R. Moore, p. 9.

Chapter 4: The Front Row: The Old Guard

Niels Bohr

My main sources for information about Niels Bohr are his own *Collected Works* (CW), the wonderful biography by Abraham Pais (1991), and the commemorative set of essays edited by Stefan Rozental. On the crucial 1926 meeting with Schrödinger, see also the essay by Jorgen Kalckar in CW, vol. 6. For a less technical biography of the man, I recommend the one by Ruth Moore.

52 *The arrival of a letter from Pauli . . . with his sardonic smile:* Leon Rosenfeld essay in Rozental, p. 114.

53 *At each place . . . to explain later:* Richard Courant essay in Rozental, p. 303.

54 *Only one of us is allowed . . . day after day in agony:* Weisskopf, p. 70.

55 *The discussion . . . with Bohr and with Schrödinger:* Heisenberg (1971), p. 73ff.

56 *If all this . . . grateful that you did:* Heisenberg (1971), p. 75.

56 *I don't mean to criticize, but:* See, for example, Dirac in C. Weiner, p. 136. Note also that this motto is shown at the bottom of Gamow's illustration of the institute in Gamow (1966), p. 165.

Paul Ehrenfest

The main source for information about Paul Ehrenfest is Martin J. Klein's wonderfully informative biography of him. Unfortunately it takes us only up to 1920. Almost all quotations in this section are taken from that book. The letters from Einstein to Ehrenfest may also be seen in *The Collected Papers of Albert Einstein* (1987–) published by the Princeton University Press.

58 *Isn't the feeling . . . How to cure it:* Ehrenfest diary, quoted in Klein, p. 47.

59 *Within a few hours . . . meant for each other:* Klein, p. 177, and p. 215 of essay "Paul Ehrenfest" by Einstein (1950).

59 *I am frankly annoyed . . . for a little while:* Letter from Einstein to Ehrenfest, May 12, 1912, quoted in Klein, p. 180.

60 *He lectures . . . in an extraordinary manner:* Letter from Sommerfeld to Lorentz, May 24, 1912, quoted in Klein, p. 184.

61 *I have never been . . . I belong to you:* Letter from Einstein to Ehrenfest, November 9, 1919, quoted in Klein, p. 314.

62 *She found Ehrenfest's need . . . close acquaintance:* Letter from Meitner to Klein, February 12, 1958, quoted in Klein, p. 49.

62 *You had gone, the music had faded away:* Letter from Ehrenfest to Bohr, June 4, 1919, in CW, vol. 3, p. 16.

62 *What I can do . . . done by others:* Letter from Ehrenfest to Einstein, August 6, 1920, quoted in Klein, p. 319.

62 *Don't complain . . . the conscience of others:* Letter from Einstein to Ehrenfest, August 13, 1920, quoted in Klein, p. 319.

62 *Don't be impatient . . . afraid of being squashed:* Letter from Ehrenfest to Einstein, September 2, 1920, quoted in Klein, p. 319.

Lise Meitner

The main source for information about Lise Meitner is Ruth Lewin Sime's extremely thorough and fascinating biography of her.

66 *He had an unusually pure . . . for his own person:* Meitner quoted by Sime, p. 406.

67 *Were I asked . . . I have ever known:* Pais (1982), p. vii.

67 *Amazons are abnormal . . . especially in the next generation:* Sime, p. 26.

68 *All these people . . . in those circles:* Hahn, p. 91.

68 *I never had time for it:* Sime, p. 35.

69 *Not often in life . . . smiling and explaining; It was one of the greatest experiences . . . I shall never forget our talks:* This exchange between Bohr and Einstein is narrated in Pais (1991), p. 228. These letters, from Einstein to Bohr, May 2, 1920, and Bohr to Einstein, June 24, 1920, are published in CW, vol. 3, pp. 22 and 634. The first of them is also in Calaprice, p. 73.

Chapter 5: The Front Row: The Revolutionaries

Werner Heisenberg

David Cassidy has written, in my opinion, the definitive biography of Heisenberg. This book has been my main source for the life of this complex and extraordinary man.

75 *Dear Harald: . . . longing to talk with you:* Letter from Bohr to his brother, Harald, June 19, 1912, in CW, vol. 1, p. 559.

75 *The single life . . . person next to me:* Letter from Heisenberg to his mother, November 12, 1936, quoted in Cassidy, p. 367.

75 *I never saw my father . . . forthcoming than that:* Quoted by Heisenberg's son Jochen in Frayn.

76 *It must have been . . . a new star to steer by:* Heisenberg (1971), p. 1.

77 *My first two years . . . between the two:* Heisenberg (1971), p. 27.

78 *I remember . . . transferred itself to others:* Letter from Heisenberg to his mother, December 15, 1930, quoted in Cassidy, p. 289.

Wolfgang Pauli

We are very fortunate to now have the biography of Pauli by Charles Enz, Pauli's last assistant and himself a distinguished professor of physics.

79 *Bohr's scientific correspondence . . . deep friendship:* Jorgen Kalckar, CW, vol. 6, p. vi.

80 *Herr Pauli . . . just the opposite:* Enz, p. 89.

81 *I've found . . . they are feminine:* Letter from Pauli to Gregor Wentzel, December 5, 1926, in PSC, vol. 1, p. 360.

81 *Alone in the express train . . . my great neurosis:* Meier, p. 151.

81 *In case my wife runs away . . . printed notice:* Letter from Pauli to Oskar Klein, February 10, 1930, in PSC, vol. 2, p. 2.

81 *Had she taken . . . an ordinary chemist:* Pauli (1994), p. 18.

81 *The underlying physical laws . . . are thus completely known:* Dirac (1929), p. 714.

Paul Dirac

For further information about Dirac, I recommend Helge Kragh's excellent biography (1990) and the volume edited by Peter Goddard.

83 *as shy as a gazelle . . . as a Victorian maid: Sunday Dispatch,* November 19, 1933; cf. also Goddard, p. 20.

83 *How can you do . . . the exact opposite:* Pais essay in Goddard, p. 51.

84 *Of all physicists, Dirac has the purest soul:* Rudolf Peierls quotes Bohr in Taylor, p. 5.

85 *My father . . . that started very early:* Dirac (1980), quoted in Schweber, p. 17; see also Margit Dirac's section in Kursonoglu and Wigner.

86 *While I was . . . very seldom provided such statements:* C. Weiner, p. 116.

86 *This was—I remember . . . perfect in its way and admirable:* Born (1968), p. 226.

Classical Mechanics versus Quantum Mechanics

89 *Even for the simplest . . . observable quantities occur:* Van der Waerden, p. 262.

91 *Only one basic . . . certain quantum conditions:* Dirac (1926), p. 561.

92 *I admired Bohr . . . practically all the talking:* C. Weiner, p. 134.

93 *We were on the steamer . . . the girls are nice:* Heisenberg in Mehra, p. 816.

93 *Heisenberg climbed . . . a tragic result:* Dirac (1971), quoted in Schweber, p. 21.

94 *Every Ph.D thesis . . . for quantum mechanics:* Weisskopf, p. 67.

Chapter 6: The Front Row: The Young Ones

***The Curse of the* Knabenphysik**

96 *There it goes, Mike. Kiss it good-bye:* This story also appears in Brown and Rigden, p. 131.

Max Delbrück

The biography of Delbrück by Ernest Fischer and Carol Lipson, though primarily concerned with his scientific career as a biologist, emphasizes his deep connection to the notion of complementarity and to Bohr personally. This link is also emphasized in Gunther Stent's introductory essay to Delbrück's *Mind from Matter?*

99 *I have accepted . . . entertaining friendly relations:* Letter from Delbrück to Bohr, June 1932, quoted in Fischer, p. 60.

Chapter 7: The Coming Storm

The Periodic Table

102 *To have perceived . . . to read the mind of God:* Sacks, p. 191.

102 *The underlying physical laws . . . thus completely known:* Dirac (1929), p. 714.

106 *It is as though . . . and for the harmonious:* Pauli (1994), p. 59.

106 *The process of understanding . . . and their behavior:* Pauli (1994), p. 221.

106 *To see and handle . . . in terms of abstract mathematics:* Fischer and Lipson, p. 26.

The New Kepler

107 *He was an incessant worker . . . came much later:* Nielsen, p. 22.

109 *One thought spectra are marvelous . . . the way to look at it:* Niels Bohr, interview by T. S. Kuhn, L. Rosenfeld, E. Rudinger, and A. Petersen, October 31, 1962, AHQP.

110 *This is an enormous . . . must then be right:* Letter from Hevesy to Bohr, September 23, 1913, CW, vol. 2, p. 532.

110 *Bohr's work . . . has driven me to despair:* Letter from Ehrenfest to Lorentz, August 25, 1913, quoted in Klein, p. 278.

111 *In Copenhagen, they could quantize your grandmother:* Stanley Deser, in Lindstrom, p. 49; see also Pais (2000), p. 138.

111 *What we are hearing . . . of Bohr for all time:* Sommerfeld (1923a), p. 1.

112 *As long as German science . . . the ranks of civilized nations:* Heilbron, p. 88n3.

Göttingen in 1922

113 *Even though I consider it . . . success along this path:* Letter from Ehrenfest to Sommerfeld, May 1916, quoted in Klein, p. 286.

113 *I have read . . . how you saw it all:* Letter from Ehrenfest to Bohr, September 27, 1921, in CW, vol. 3, p. 627.

114 *It is interesting . . . of a chemical nature:* Kragh (1979), p. 156, and French and Kennedy, p. 60.

114 *a somewhat mystical . . . outside Copenhagen:* Kragh (1979), p. 156, and French and Kennedy, p. 60.

115 *We had all of us . . . from Bohr's own lips:* Heisenberg (1971), p. 38.

115 *insight into the structure . . . inexorable test of experiment:* Heisenberg essay in Rozental, p. 95.

116 *This walk . . . only began that afternoon:* Heisenberg (1971), p. 38.

116 *It was my first conversation . . . worried by its difficulties:* Werner Heisenberg, interview by T. S. Kuhn, November 30, 1962, AHQP.

117 *After a few days . . . with our rucksacks:* Heisenberg (1971), p. 46.

117 A *new phase of my scientific life . . . the first time:* Pauli (1946).

Triumph and Crisis

118 *The only thing I know . . . to be a rare earth:* Letter from Bohr to James Franck, July 15, 1922, in CW, vol. 4, p. 24.

119 *considered at the awarding . . . same time as you:* Letter from Bohr to Einstein, November 11, 1922, in CW, vol. 4, p. 28.

119 *I can say . . . fondness of your mind even greater:* Letter from Einstein to Bohr, January 10, 1923, in CW, vol. 4, p. 28.

120 *to the vigorous growth . . . sorrowful times:* CW, vol. 4, p. 27.

121 *All attempts . . . have proved to be unsuccessful:* Sommerfeld (1923b).

121 *an old acquaintance . . . not a respectable fellow:* Letter from Einstein to Ehrenfest, May 31, 1924, Einstein Archives, Hebrew University of Jerusalem.

121 *can't be done without:* Letter from Einstein to Ehrenfest, July 12, 1924, Einstein Archives, Hebrew University of Jerusalem.

122 *an employee in a gaming house than a physicist:* Letter from Einstein to Born, April 29, 1924, quoted in Born (1971), p. 82.

122 *Even if it were . . . authorities are very contradictory:* Letter from Pauli to Bohr, October 2, 1924, in PSC, vol. 1, p. 163.

122 *There is nothing . . . as honorable a funeral as possible:* Letter from Bohr to R. Fowler, April 21, 1925, in CW, vol. 5, p. 81.

122 *was very comforting . . . doubt the energy principle:* Letter from Bohr to S. Rosseland, January 6, 1926, in CW, vol. 5, p. 484.

The New Optimism

124 *Dear Pauli! . . . Merry Christmas— Werner Heisenberg:* Postcard from Heisenberg to Pauli, December 15, 1924, in PSC, vol. 1, p. 142.

125 *My conscience is so bad . . . to quarrel with you again:* Letter from Bohr to Pauli, December 11, 1924, in CW, vol. 5, p. 34.

125 *So you see . . . at a crucial turning point:* Letter from Bohr to Pauli, December 22, 1924, in PSC, vol. 1, p. 193.

126 *Well, you are both young, you can afford a stupidity:* George Uhlenbeck, interview by T. Kuhn, March 31, 1962, AHQP.

127 *Nominate . . . stop Albert Einstein:* Telegram from Einstein to the Royal Swedish Academy of Sciences, January 13, 1945, quoted in Pais (1982), p. 517; copy also in Einstein Archives, Hebrew University of Jerusalem.

127 *To Wolfgang Pauli . . . also named Pauli Principle: Nobel Lectures* (1964).

Chapter 8: The Revolution Begins

Helgoland

129 *energy and momentum . . . is an example:* Letter from Pauli to Bohr, December 12, 1924, in CW, vol. 5, p. 426; see also PSC, vol. 1, p. 186.

130 *Weak men . . . collide with one another:* Letter from Pauli to Bohr, December 31, 1924, in PSC, vol. 1, p. 197.

130 *When I was young . . . not a revolutionary:* Mehra and Rechenberg, vol. 1, p. xxiv, and Pais (1986), p. 314.

130 *He does not . . . further our science:* Letter from Pauli to Bohr, February 11, 1924, in PSC, vol. 1, p. 143.

131 *There was a moment . . . and was happy:* Heisenberg in Van der Waerden, p. 25.

132 *convinced in his heart . . . is truly right:* Letter from Heisenberg to Pauli, June 29, 1925, in PSC, vol. 1, p. 229.

132 *It is well known . . . more or less fortuitous:* Heisenberg (1925), translated in Van der Waerden, p. 261.

Another Sleepless Night

133 *I began to ponder . . . since my student days:* Born in Van der Waerden, p. 37, and Richter, p. 60.

134 *I joined him . . . by your futile mathematics:* Van der Waerden, p. 37.

134 *One has to ensure . . . deluge of Göttingen erudition:* Letter from Pauli to Ralph Kronig, October 9, 1925, in PSC, vol. 1, p. 242.

135 *Heisenberg's form . . . motion of electrons:* Pauli (1926), translated in Van der Waerden, p. 387.

135 *I don't need to write . . . so quickly:* Letter from Heisenberg to Pauli, November 3, 1925, in PSC, vol. 1, p. 252.

135 *To my great joy . . . about it very soon:* Letter from Bohr to Pauli, November 13, 1925, in PSC, vol. 1, p. 254.

Waves or Particles

137 *I believe . . . our physics enigmas:* Letter from Einstein to Lorentz, December 16, 1924, Einstein Archives, Hebrew University of Jerusalem.

137 *When quite young . . . centuries with honor:* 1929 Nobel Prize in Physics Presentation Speech, *Nobel Lectures* (1965).

138 *After long reflections . . . notably to electrons:* de Broglie, p. 4.

140 *Schrödinger did . . . outburst in his life:* Remark by Hermann Weyl to Abraham Pais, Pais (1986), p. 252.

Heisenberg versus Schrödinger

141 *the way an inquisitive child . . . long time:* Letter from Planck to Schrödinger, February 4, 1926, in Przibram, p. 3.

141 *The idea . . . a real genius:* Letter from Einstein to Schrödinger, April 16, 1926, in Przibram, p. 24.

142 *I am convinced . . . off the track:* Letter from Einstein to Schrödinger, April 26, 1926, in Przibram, p. 28.

142 *every day . . . in all the splendid ramifications:* Letter from Ehrenfest to Schrödinger, May 29, 1926, quoted in W. Moore, p. 491.

142 *Young man . . . nonsense about quantum jumps:* Heisenberg (1971), p. 73, and letter from Heisenberg to Pauli, July 28, 1926, in PSC, vol. 1, p. 337.

143 *My theory was inspired . . . if not to say repelled:* Schrödinger footnote in W. Moore, p. 209.

144 *Quantum mechanics . . . He does not play dice:* Letter from Einstein to Born, December 4, 1926, in Born (1971), p. 90.

145 *After 8 or 9 o'clock . . . until twelve or one o'clock at night:* Werner Heisenberg, interview by T. Kuhn, February 19, 1963, AHQP.

146 *You no longer . . . he is defending:* Letter from Schrödinger to Wilhelm Wien, October 21, 1926, quoted in W. Moore, p. 228.

146 *Most other . . . at the wall:* Niels Bohr, interview by T. Kuhn, February 19, 1963, AHQP.

146 *One may view the world . . . and make rude remarks:* Letter from Pauli to Heisenberg, October 19, 1926, in PSC, vol. 1, p. 347.

148 *Do not enter this . . . in that juggernaut:* Undated memorandum of Kramers to Klein, NBA, quoted in Pais (2000), p. 159.

Uncertainty and Complementarity

148 *Both of us . . . complicated problems:* Heisenberg (1971), p. 77.

149 *I know very well . . . await your merciless criticism:* Letter from Heisenberg to Pauli, February 23, 1927, in PSC, vol. 1, p. 376.

149 *Bohr tried to explain . . . pressure from Bohr*: Werner Heisenberg, interview by T. Kuhn, February 19, 1963, AHQP.

149 *Bohr and I talked . . . that one is convinced of*: Pais (1991), p. 239.

150 *In the sharp formulation . . . elements of the present*: Heisenberg (1927).

152 *Because atomic behavior . . . direct human experience*: Feynman, section 37-1.

Chapter 9: The King in Decline

The Crucial Solvay Conference

156 *If I get to Paris . . . any more either*: Berg, p. 115.

157 No *more profound . . . that is the way to do it*: Snow (1981), p. 111.

157 *Einstein was so incredibly sweet . . . humane and friendly*: Niels Bohr, interview by A. Bohr and L. Rosenfeld, July 12, 1961, NBA, quoted by Pais (1991), p. 229.

157 *I have thought a hundred times . . . Relativity Theory*: Einstein remark to Otto Stern, quoted in Pais (1982), p. 9.

158 *Bohr towering . . . contra Einstein:* Letter from Ehrenfest to Goudsmit, Uhlenbeck, and Dieke, November 3, 1927, in CW, vol. 6, p. 38.

159 *In these discussions . . . examinations in front of them:* Holton and Elkana, p. 84.

160 *It made me so happy . . . as in the "old days"*: Letter from Heisenberg to Bohr, October 2, 1928, NBA, quoted in Pais (1991), p. 309.

160 *Thanking you . . . any other human being:* Letter from Bohr to Heisenberg, undated (December 1928), in CW, vol. 6, p. 24.

Einstein—The King

161 *We British physicists . . . our hopes will be fulfilled:* Letter from Pauli to Bohr, June 16, 1928, in PSC, vol. 1, p. 462; also translated in CW, vol. 6, p. 52.

161 *The Heisenberg-Bohr . . . let him lie there*: Letter from Einstein to Schrödinger, May 31, 1928, in Przibram, p. 31.

161 *Many of us regard . . . leader and standard-bearer*: Quoted by M. J. Klein in French, p. 150.

162 *for the time being . . . sorry to see it happen:* Letter from Pauli to Einstein, December 19, 1929, in PSC, vol. 1, p. 526.

162 *I would like to add . . . switched to pure mathematics:* Letter from Pauli to Einstein, December 19, 1929, in PSC, vol. 1, p. 526.

162 *I found your letter very amusing . . . forces of Nature:* Letter from Einstein to Pauli, December 24, 1929, in PSC, vol. 1, p. 528; also in Einstein Archives, Hebrew University of Jerusalem.

162 *Study the problem . . . what you think about it:* Letter from Einstein to Pauli, December 24, 1929, in PSC, vol. 1, p. 528; also in Einstein Archives, Hebrew University of Jerusalem.

162 *Dear Pauli: You were right, you* Spitzbube: Letter from Einstein to Pauli, January 22, 1932, in PSC, vol. 2, p. 109; also in Einstein Archives, Hebrew University of Jerusalem.

Chapter 10: The Great Synthesis

Dirac's Equation

165 *I was more interested in getting the correct equations:* Holton and Elkana, p. 84.

166 *Schrödinger and I . . . the basis of much of our success:* C. Weiner, p. 136.

166 *I was really scared . . . new domains:* C. Weiner, p. 143.

167 *Schrödinger and Dirac . . . gladly shared with Born:* Letter from Heisenberg to Bohr, November 27, 1933, quoted in Cassidy, p. 325.

How Max Delbrück Joined Knabenphysik

169 *I found out . . . felt right at home:* quoted in Fischer and Lipson, p. 38, and Beranek, p. 528.

171 *It was very difficult . . . to go more slowly afterwards:* Born (1968), p. 37.

171 *The underlying physical laws . . . are thus completely known:* Dirac (1929), p. 714.

172 *acceptable but rather dull:* Fischer and Lipson, p. 50.

Physics Begins to Split

173 *Thus disappeared . . . universal excellence:* Segrè (1980), p. 222.

175 *the strongest emotional experience . . . Nature had spoken to him:* Pais (1982), p. 253.

175 *for a few days . . . with joyous excitement:* Letter from Einstein to Ehrenfest, January 17, 1915, as quoted in Pais (1982), p. 253.

176 *a person first . . . anxiety to him:* Paul Dirac, interview by T. Kuhn, May 7, 1963, AHQP.

176 *I have completely lost . . . I cannot be helped any more:* Letter from Ehrenfest to Bohr, May 13, 1931, NBA, as quoted in Pais (1991), p. 418.

Delbrück's Choices

178 *The results . . . the lighter atoms:* Rutherford (1919), p. 581.

179 *They play games . . . the real facts of nature:* Blackett, p. 61.

Chapter 11: Conservation of Energy

The Mysteries of the Nucleus

182 *You shouldn't send . . . which I drink on Sundays:* Letter from Lise Meitner to Hedwig Meitner, May 4, 1923, as quoted in Sime, p. 98.

The Barrier Is Too High

184 *I shall never forget . . . even in his physics:* Quoted by Barry Streuwer in Harper, p. 33.

185 *Before I closed the magazine . . . no impenetrable barriers:* Gamow (1970), p. 60.

186 *Anyone present in this room . . . out through the window:* Gamow (1970), p. 60.

187 *My secretary tells me . . . here for one year:* Gamow (1970), p. 59.

188 *I thought you . . . And so they did:* Gamow (1970), p. 83.

189 *Landau's was . . . ever came across:* Casimir, p. 106.

189 *We formed a trio . . . appreciated by others:* Casimir, p. 105.

189 *arguing and gesticulating . . . to convince him he was wrong:* Frisch, p. 101.

Heaven and Earth

192 *There is an occasional correct sentence:* Casimir, p. 117.

192 *It has never been . . . will undoubtedly do:* Casimir, p. 117.

194 *helplessness in the discussion of experiments:* Letter from Pauli to Klein, February 18, 1929, in PSC, vol. 1, p. 488.

194 *Do you intend to mistreat the poor energy law further:* Letter from Pauli to Bohr, March 5, 1929, in PSC, vol. 1, p. 493.

194 *let the stars shine in peace:* Letter from Pauli to Bohr, July 17, 1929, in PSC, vol. 1, p. 512.

194 *Fraulein Meitner . . . Copenhagen theoretical nonsense:* Letter from Pauli to Klein, March 16, 1929, in PSC, vol. 1, p. 494.

194 *Let this note rest for a good long time:* Letter from Pauli to Bohr, July 17, 1929, in PSC, vol. 1, p. 512.

195 *My own opinion . . . conservation of energy:* Letter from Dirac to Bohr, November 26, 1929, NBA.

195 *I have heard . . . in our philosophy:* Letter from Rutherford to Bohr, November 19, 1929, NBA.

The Revolutionary Proposal

197 *There's not the smallest orb . . . we cannot hear it:* William Shakespeare, *The Merchant of Venice,* act 5, scene 1.

198 *We really do not . . . we understand nothing:* Letter from Pauli to Bohr, July 17, 1929, in PSC, vol. 1, p. 513.

198 *In real life . . . people's resistance:* Meier, p. 135.

198 *Dear Radioactive . . . humble servant, Wolfgang Pauli:* Letter from Pauli to Meitner, December 4, 1930, in Pauli (1994), p. 198.

200 *that foolishly behaving foolish child of my life's 1930–31 crisis:* Letter from Pauli to Delbrück, October 6, 1958, Pauli Archive, CERN Scientific Information Service, Geneva, Switzerland, as quoted by Enz, p. 533.

201 *small bourgeois, Philistine . . . coffee . . . alcohol:* Letter from Pauli to Gregor Wentzel, September 7, 1931, in PSC, vol. 3, p. 752.

The Three Young Geniuses Each Write a Book

202 *I have given up . . . too difficult for me:* Letter from Heisenberg to Bohr, July 27, 1931, Bohr Scientific Correspondence, vol. 20, sec. 2, AHQP, quoted in Cassidy, p. 290.

204 *not quite as good . . . of quantum mechanics:* Casimir, p. 51

204 *It has become . . . without introducing irrelevancies:* Dirac (1958), p. vii.

205 *The new theories . . . their properties and uses:* Dirac (1958), p. vii.

205 *doesn't provide . . . you didn't have before:* Paul Dirac, interview by T. Kuhn, May 7, 1963, AHQP.

Chapter 12: The New Generation Comes of Age

The Apprenticeship

208 *a very bad quality . . . definitive truth:* Letter from Pauli to Ehrenfest, February 15, 1929, in PSC, vol. 1, p. 486.

209 *One thing . . . The Scourge of God:* Letter from Pauli to Ehrenfest, February 15, 1929, in PSC, vol. 1, p. 486.

211 *One of my tasks . . . in the afternoon:* R. Kronig, "The Turning Point," in Fierz and Weisskopf, pp. 5–39.

211 *would insist . . . half-baked argument:* Peierls, p. 48.

211 *You often realized . . . turned a corner:* Peierls, p. 121

212 *When he was faced . . . risky for others to imitate:* Peierls, p. 33.

212 *His methods were always simple . . . became very simple:* Peierls, p. 88.

Copenhagen 1932

214 *Do you also belong . . . lack of respect seriously either:* Bloch, p. 32.

216 *Yet the concentration . . . with the real work:* Casimir, p. 125.

216 *In a Western picture . . . question the wisdom of Bohr:* Casimir, p. 97.

The "Blegdamsvej Faust"

219 *Only the friendship . . . otherwise foreign territory:* Letter from Heisenberg to his mother, October 8, 1934, as quoted in Cassidy, p. 326.

Delbrück's Dilemma

227 *Thanks for the message . . . knows how to wait:* Quoted in Enz, p. 489.

Chapter 13: The Miracle Year

The Discovery of the Neutron

231 *As I told him . . . I recall no similar occasion:* Chadwick, 1964, in the *Proceedings of the Tenth Annual Congress on the History of Science,* as quoted by Rhodes, p. 162.

232 *If we suppose . . . in his Bakerian lecture of 1920:* Chadwick (1932a), p. 312.

232 *Now I want . . . put to bed for a fortnight:* Snow (1981), p. 35.

232 *For the neutron . . . deserve it for something else:* Segrè (1980), p. 184.

Copenhagen and the Neutron

234 *A great truth . . . also a great truth:* French and Kennedy, p. 223.

234 *The possibility that . . . of such nuclei as nitrogen-14:* Chadwick (1932b), p. 392.

234 *When he was faced . . . give him the answer:* Peierls, p. 33.

235 *equal to zero . . . the electron mass:* E. Fermi, vol. 1, p. 568.

236 *The conjectures . . . physical reality:* Segrè (1970), p. 75.

The Miracle Year

237 *When the behavior . . . against its validity:* Blackett and Occhialini, p. 699.

237 *The discovery . . . our whole picture of matter:* Mehra (1973), p. 271.

238 *family-theorist:* Letter from Delbrück to Bohr, June 1932, quoted in Fischer, p. 60.

239 *Finally, it may be remarked . . . developed in Fermi's theory:* Bohr (1936), p. 25.

240 *the prehistory . . . of nuclear physics:* Hans Bethe, oral history interview on record at the American Institute of Physics, p. 3, as quoted by Rhodes, p. 165.

240 *Progress in the field . . . will bring:* Eve, p. 356.

The Hammer and the Needle

244 *To those who look . . . are the merest moonshine:* Rutherford (1933), p. 432.

245 *I congratulate you . . . the sphere of theoretical physics:* Letter from Rutherford to Fermi, April 23, 1934, in E. Fermi, vol. 1, p. 841.

247 *civil liberty . . . equality before the law:* See Clark, p. 264, and Calaprice, p. 106.

Chapter 14: Ehrenfest's End

251 *O Caasje . . . the wagon of Leiden physics:* Casimir, p. 147.

252 *What you have just said . . . the force to live:* Letter from Dirac to Bohr, September 28, 1933, NBA, as quoted in Pais (1991), p. 410.

253 *a man of scintillating intellect . . . insufficiently regarded:* Pauli (1933), p. 884, and Pauli (1994), p. 79.

Epilogue: Or What Happened Afterward to the Front Row's Other Six

How Meitner Discovered Nuclear Fission

258 *There is something . . . telling only you:* Letter from Hahn to Meitner, December 12, 1938, Meitner Papers, Churchill College, Cambridge University, quoted by Sime, p. 233.

258 *Oh, what idiots . . . as it must be:* Frisch, p. 116.

How Bohr Lived Happily Ever After

260 *The temptation is . . . candles along:* Letter from Pauli to Bohr, July 17, 1929, in PSC, vol. 1, p. 512.

260 *How nice it would be . . . for a long time:* Letter from Bohr to Pauli, July 31, 1929, in CW, vol. 6, p. 194.

How Dirac Got Married

262 *My own contributions . . . of minor importance:* Dirac (1949), p. 392.

How Heisenberg Inspired His Friend to Paint Like Titian

264 *This is to show . . . details are missing:* Pauli, as quoted by Enz, p. 530.

How Pauli's Anima Made Him Leave the United States

265 *I will never forget . . . as his successor:* Letter from Pauli to Born, April 25, 1955, in PSC, vol. 3, p. 412.

265 *As in Austria . . . departed the* U.S.A.: Letter from Pauli to M. L. Franz, May 17, 1951, in PSC, vol. 3, p. 306.

How Delbrück Became a Biologist

268 *Max's personal integrity . . . the scientific life:* James D. Watson, quoted on Fischer and Lipson book cover.

268 *Max took me . . . contributed so little:* Luria, p. 33.

268 *I cannot forget . . . that is important:* Letter from Pauli to Delbrück, October 6, 1958, Pauli Archive, CERN Scientific Information Service, Geneva, Switzerland, as quoted by Enz, p. 533.

269 *This was the worst . . . a word of it:* Quoted by J. Weiner, p. 47.

269 *God and Mephisto, all in one:* Perutz, p. 180.

Bibliography

Bell, John S. 1987. *Speakable and Unspeakable in Quantum Mechanics*. Cambridge: Cambridge University Press.

Beranek, W., ed. 1972. *Science, Scientists and Society*. New York: Bogden and Quigley.

Berg, A. Scott. 1998. *Lindbergh*. New York: G. P. Putnam's Sons.

Bernstein, Jeremy. 2004. *J. Robert Oppenheimer: Portrait of an Enigma*. Chicago: Ivan Dee.

Bird, Kai, and Martin J. Sherwin. 2005. *American Prometheus. The Triumph and Tragedy of J. Robert Oppenheimer*. New York: Knopf.

Blackett, Patrick M. 1972. "Rutherford." *Notes and Records of the Royal Society of London* 27:57–72.

Blackett, Patrick M., and Giuseppe Occhialini. 1933. "Some Photographs of the Tracks of Penetrating Radiation." *Proceedings of the Royal Society of London* A 139:699–720.

Bloch, Felix. 1963. "Reminiscences of Niels Bohr." *Physics Today*, October, p. 32.

Blumberg, Stanley, and Gwinn Owens. 1976. *Energy and Conflict: The Life and Times of Edward Teller*. New York: G. P. Putnam's Sons.

Bohr, Niels. 1936. "Conservation Laws in Quantum Theory." *Nature* 138:25. Reprinted in *Collected Works*, vol. 5, p. 213.

———. 1972– . *Collected Works*. 11 volumes. Amsterdam: North-Holland. Includes some correspondence and commentary as well.

Boorstin, Daniel. 1992. *The Creators*. New York: Random House.

Born, Max. 1968. *My Life and My Views*. New York: Scribners.

———. 1971. *The Born-Einstein Letters: Friendship, Politics and Physics in Uncertain Times*. Trans. Irene Born. London: Macmillan.

Brown, Laurie. 1978. "The Idea of the Neutrino." *Physics Today*, September, pp. 23–28.

Brown, Laurie, Abraham Pais, and Brian Pippard, eds. 1995. *Twentieth Century Physics*. 3 vols. New York: American Institute of Physics Press.

Brown, Laurie, and John Rigden, eds. 1993. *Most of the Good Stuff: Memories of Richard Feynman*. New York: American Institute of Physics.

Calaprice, Alice. 2005. *The New Quotable Einstein*. Princeton, N.J.: Princeton University Press.

Canady, John. 2000. *The Nuclear Muse*. Madison: University of Wisconsin Press.

Casimir, Hendrik. 1983. *Haphazard Reality: Half a Century of Physics*. New York: Harper and Row.

Cassidy, David. 1992. *Uncertainty: The Life and Science of Werner Heisenberg*. New York: W. H. Freeman.

Cathcart, Brian. 2005. *The Fly in the Cathedral*. New York: Farrar, Straus and Giroux.

Chadwick, James. 1932a. "Possible Existence of the Neutron." *Nature* 129:312–13.
——. 1932b. "The Existence of the Neutron." *Proceedings of the Royal Society of London* A 136:392.
Clark, Ronald. 1984. *The Life and Times of Einstein.* New York: Harry Abrams.
Crease, Robert, and Charles Mann. 1986. *The Second Creation: Makers of the Revolution in Twentieth-Century Physics.* New York: Collier.
Curie, Irène and Frédéric Joliot. 1932. *Comptes Rendus de l'Académie des Sciences* 194:273.
de Broglie, Louis. 1963. *Recherches sur la théorie des quanta.* Paris: Masson.
Delbrück, Max. 1986. *Mind from Matter?* Palo Alto, Calif.: Blackwell.
Dirac, Paul. 1926. "Quantum Mechanics and a Preliminary Investigation of the Hydrogen Atom." *Proceedings of the Royal Society of London* A 110:561–79.
——. 1929. "The Quantum Mechanics of Many Electron Systems." *Proceedings of the Royal Society of London* A 123:714–733.
——. 1949. "Forms of Relativistic Dynamics." *Reviews of Modern Physics* 21:392.
——. 1958. *Quantum Mechanics,* 4th ed. Oxford: Oxford University Press.
——. 1971. *The Development of Quantum Mechanics.* New York: Gordon and Breach.
——. 1980. "A Little Prehistory." *The Old Cathamian, Annual Journal of the Merchant Venturers School,* p. 9.
Dresden, Max. 1987. *H. A. Kramers: Between Tradition and Revolution.* New York: Springer.
Einstein, Albert. 1922. *Naturwissenschaften* 10:184 (Einstein's review of Pauli's 1922 article on relativity theory). Reprinted in Wolfgang Pauli, *Collected Scientific Papers,* Ed. K. v. Meyenn with the assistance of A. Hermann and V. Weisskopf. New York: Springer, 1964.
——. 1950. *Out of My Later Years.* New York: Philosophical Library.
——. 1987. *The Collected Papers of Albert Einstein.* 9 vols. Princeton, N.J.: Princeton University Press. The first 9 volumes of a projected 25-volume set; they include writings and correspondence through 1920.
Enz, Charles. 2002. *No Time to be Brief: A Scientific Biography of Wolfgang Pauli.* New York: Oxford University Press.
Eve, Arthur S. 1939. *Rutherford.* Cambridge: Cambridge University Press.
Fermi, Enrico. 1965. *Collected Works.* 2 vols. Eds. E. Segrè et al. Chicago: University of Chicago Press.
Fermi, Laura. 1954. *Atoms in the Family.* Chicago: University of Chicago Press.
Feynman, Richard. 1965. *Lectures on Physics.* 3 vols. Reading, Mass.: Addison-Wesley.
Fierz, Marcus, and Victor Weisskopf, eds. 1960. *Theoretical Physics in the Twentieth Century: A Memorial Volume to Wolfgang Pauli.* New York: Wiley Interscience.
Fischer, Ernest, and Carol Lipson. 1988. *Thinking About Science: Max Delbrück and the Origins of Molecular Biology.* New York: W. W. Norton.
Frayn, Michael. 2002. "'Copenhagen' Revisited." *New York Review of Books,* March 28, p. 23.
French, Anthony. 1979. *Albert Einstein: A Centenary Volume.* Cambridge, Mass.: Harvard University Press.
French, Anthony, and P. J. Kennedy. 1985. *Niels Bohr: A Centenary Volume.* Cambridge, Mass.: MIT Press.

Frisch, Otto. 1979. *What Little I Remember.* Cambridge: Cambridge University Press.

Galison, Peter, and Bruce Hevly, eds. 1992. *Big Science: The Growth of Large Scale Research.* Stanford, Calif.: Stanford University Press.

Gamow, George. 1966. *Thirty Years That Shook Physics: The Story of Quantum Theory.* Reprint, New York: Dover, 1985.

——. 1970. *My World Line: An Informal Autobiography.* New York: Viking Press.

Goddard, Peter, ed. 1998. *Paul Dirac: The Man and His Work.* Cambridge: Cambridge University Press.

Goethe, Johann Wolfgang von. 1905. *Faust.* Trans. Anna Swanwick. London: G. Bell and Sons.

——. 1962. *Faust.* Trans. Walter Kaufmann. New York: Doubleday.

——. 1987. *Faust Part I.* Trans. David Luke. New York: Oxford University Press.

——. 1994. *Faust Part II.* Trans. David Luke. New York: Oxford University Press.

Gottfried, Kurt, and Tung-Mow Yan. 2005. *Quantum Mechanics: Fundamentals.* New York: Springer.

Hahn, Otto. 1970. *My Life.* Trans. Ernst Kaiser. New York: Herder and Herder.

Harper, Eamon, ed. 1997. *The George Gamow Symposium.* San Francisco: Astronomical Society of the Pacific Publications.

Heilbron, John. 1986. *The Dilemmas of an Upright Man: Max Planck as Spokesman for German Science.* Berkeley: University of California Press.

Heisenberg, Werner. 1925. "Quantum Theoretical Re-interpretation of Kinematical and Mechanical Relations." *Zeitschrift für Physik* 33:879–93. English trans. in Bartel van der Waerden, ed., *Sources of Quantum Mechanics,* p. 261. Amsterdam: North-Holland, 1967.

——. 1927. "On the Perception/Content of Quantum Theoretical Kinematics and Mechanics." *Zeitschrift für Physik* 43:172. English trans. in J. A. Wheeler and W. H. Zurek, eds., *Quantum Theory and Measurement,* p. 62. Princeton, N.J.: Princeton University Press, 1983.

——. 1971. *Physics and Beyond, Encounters and Conversations.* Trans. Arnold Pomerans. New York: Harper and Row.

——. 1974. *Across the Frontiers.* Trans. Peter Heath. New York: Harper and Row.

——. 1985– . *Gesammelte Werke.* Ed. W. Blum et al. Munich: Piper.

Hermann, Armin. 1971. *Max Planck: The Genesis of Quantum Theory.* Cambridge, Mass.: MIT Press.

Holton, Gerald, and Yehuda Elkana, eds. 1982. *Albert Einstein, Historical and Cultural Perspectives.* Princeton, N.J.: Princeton University Press.

Klein, Martin J. 1970. *Paul Ehrenfest: The Making of a Theoretical Physicist.* New York: Elsevier.

Kragh, Helge. 1979. "Niels Bohr's Second Atomic Theory." *Historical Studies in the Physical Sciences* 10:123–86.

——. 1990. *Paul Dirac: A Scientific Biography.* Cambridge: Cambridge University Press.

——. 1999. *Quantum Generations: A History in the 20th Century.* Princeton, N.J.: Princeton University Press.

Kursonoglu, Behram, and Eugene Wigner, eds. 1987. *Reminiscences About a Great Physicist: Paul Adrien Maurice Dirac.* Cambridge: Cambridge University Press.

Landau, Lev, and E. M. Lifshitz. 2003. *Quantum Mechanics: Non-Relativistic Theory.* Oxford: Butterworth Heinemann.

Lindstrom, U., ed. 1995. *The Proceedings of the Oskar Klein Centenary.* Singapore: World Scientific.

Livanova, Anna. 1980. *Landau: A Great Physicist and Teacher.* New York: Oxford University Press.

Luria, Salvador. 1984. *A Slot Machine, A Broken Test Tube.* New York: Harper and Row.

Marinetti, Filippo. 1909. "Futurist Manifesto." *Le Figaro,* Paris, February 20.

Mehra, Jagdish, ed. 1973. *The Physicist's Conception of Nature.* Dordrecht: Reidel.

Mehra, Jagdish, and Helmut Rechenberg. 1982. *The Historical Development of Quantum Theory.* 4 vols. Berlin: Springer-Verlag.

Meier, C. A., ed. 2001. *Atom and Archetype: The Pauli/Jung Letters, 1932–1958.* Trans. David Roscoe. Princeton, N.J.: Princeton University Press.

Meitner, Lise. 1958. "Max Planck als Mensch." *Naturwissenschaften* 45:406.

Meyenn, Karl v., and Charles Enz, eds. 1988. *Wolfgang Pauli: das Gewissen der Physik.* Braunschweig, Germany: Vieweg.

Meyenn, Karl von, and Engelbert Schucking. 2001. "Wolfgang Pauli." *Physics Today,* February, p. 43.

Meyenn, Karl v., Klaus Stolzenburg, and Roman Sexl, eds. 1985. *Niels Bohr, Der Kopenhagener Geist in der Physik.* Braunschweig, Germany: Vieweg.

Moore, Ruth. 1966. *Niels Bohr.* New York: Knopf.

Moore, Walter. 1989. *Schrödinger: Life and Thought.* Cambridge: Cambridge University Press.

Nielsen, J. Rud. 1962. "Memories of Niels Bohr." *Physics Today,* October, p. 22.

Nobel Lectures. 1965. *Physics 1922–1941.* Amsterdam: Elsevier.

Nobel Lectures. 1964. *Physics 1942–1962.* Amsterdam: Elsevier.

Pais, Abraham. 1982. *Subtle Is the Lord: The Science and Life of Albert Einstein.* New York: Oxford University Press.

——. 1986. *Inward Bound.* New York: Oxford University Press.

——. 1991. *Niels Bohr's Times, in Physics, Philosophy, and Polity.* New York: Oxford University Press.

——. 2000. *The Genius of Science: A Portrait Gallery of Twentieth-Century Physicists.* New York: Oxford University Press.

Pauli, Wolfgang. 1926. "On the Hydrogen Spectrum from the Standpoint of the New Quantum Mechanics." *Zeitschrift für Physik* 36:336–63. English trans. in Bartel van der Waerden, ed., *Sources of Quantum Mechanics,* p. 387. Amsterdam: North-Holland, 1967.

——. 1933. "Paul Ehrenfest." *Naturwissenschaften* 21:884.

——. 1946. "Remarks on the History of the Exclusion Principle." *Science* 103:213.

——. 1948. "Sommerfeld's Contributions to Quantum Theory." *Naturwissenschaften* 35:129.

——. 1964. *Collected Scientific Papers.* 2 vols. ed. R. Kronig and V. Weisskopf. New York: Interscience.

——. 1975. *Scientific Correspondence with Bohr, Einstein, Heisenberg, and Others.* 6 vols. Ed. K. v. Meyenn with the assistance of A. Hermann and V. Weisskopf. New York: Springer.

——. 1994. *Writings on Philosophy and Physics*. Ed. Charles Enz and Karl von Meyenn. Berlin: Springer-Verlag.
Peierls, Rudolf. 1985. *Bird of Passage*. Princeton, N.J.: Princeton University Press.
Perutz, Max. 1998. *I Wish I'd Made You Angry Earlier: Essays on Science and Scientists*. Plainview, N.Y.: Cold Spring Harbor Laboratory Press.
Przibram, Karl, ed. 1967. *Letters on Wave Mechanics*. Trans. M. J. Klein. New York: Philosophical Library.
Rhodes, Richard. 1986. *The Making of the Atomic Bomb*. New York: Simon and Schuster.
Richter, Stefan. 1979. *Wolfang Pauli: Die Jahre 1918–1930*. Aarau, Switzerland: Sauerländer.
Rosenfeld, Leon. 1966. *Nuclear Structure with Neutrons*. Amsterdam: North-Holland.
Rozental, Stefan, ed. 1967. *Niels Bohr: His Life and Work as Seen by His Friends and Colleagues*. Amsterdam: North Holland.
Rutherford, Ernest. 1919. "Collisions of Alpha Particles with Light Atoms." *Philosophical Magazine* 37:581.
——. 1933. *Nature* 132:432–33. Summary of a speech Rutherford gave at the British Association for the Advancement of Science on September 11, 1933.
Sacks, Oliver. 2001. *Uncle Tungsten: Memories of a Chemical Boyhood*. New York: Knopf.
Schrödinger, Erwin. 1926. "Quantization as an Eigenvalue Problem." *Annalen der Physik* 79:734–56.
Schweber, Silvan. 1994. *QED and the Men Who Made It*. Princeton, N.J.: Princeton University Press.
Segrè, Emilio. 1970. *Enrico Fermi: Physicist*. Chicago: University of Chicago Press.
——. 1980. *From X-Rays to Quarks*. San Francisco: W. H. Freeman.
——. 1993. *A Mind Always in Motion*. Berkeley: University of California Press.
Sime, Ruth. 1996. *Lise Meitner: A Life in Physics*. Berkeley: University of California Press.
Snow, Charles. 1967. *Variety of Men*. New York: Scribners.
——. 1981. *The Physicists*. Boston: Little, Brown.
Sommerfeld, Arnold. 1923a. *Atomic Structure and Spectral Lines*. New York: Dutton.
——. 1923b. *Reviews of Scientific Instruments* 7:509.
Taylor, John C., ed. 1987. *Tributes to Paul Dirac*. Bristol, U.K.: Adam Hilger.
Trigg, George. 1995. *Landmark Experiments in Twentieth Century Physics*. New York: Dover.
van der Waerden, Bartel, ed. 1967. *Sources of Quantum Mechanics*. Amsterdam: North-Holland.
Weiner, Charles, ed. 1977. *History of Twentieth Century Physics: Proceedings of the International School of Physics "Enrico Fermi,"* Course LVII. New York: Academic Press.
Weiner, Jonathan. 1999. *Time, Love, Memory: A Great Biologist and His Quest for the Origins of Behavior*. New York: Knopf.
Weisskopf, Victor. 1991. *The Joy of Insight: Passions of a Physicist*. New York: Basic Books.
Wilson, David. 1983. *Rutherford, Simple Genius*. London: Hodder.

Index